# COSMIC

# PHENOMENA

# GABRIELE VANIN

# COSMIC
# PHENOMENA

## FIREFLY BOOKS

*The author wishes to thank all those who have furnished graphic material for the book and, in particular, the Italian astrophiles who have contributed so much so generously, demonstrating once again the high level reached by astrophotography in this country.*

*Special thanks go to the following people: Fred Espenak of NASA's Goddard Space Flight Center and Jay Anderson of Environment Canada for their timely provision of NASA's bulletin about the August 11, 1999, eclipse, which serves as the basis for Chapter 11; Claudio Costa, director of the Research Division of the Italian Astrophiles' Union for his computer research on eclipses in history; Enrico Stomeo, head of the Meteor Division of the Italian Astrophiles' Union for his help in researching images; Claudia Amerio of Genoa for calculating the formula that underlies the table on page 144; Enzo Garberoglio from Belluno for calling my attention to the citation on page 24; Gianvittore Delaito of the Associazione Astronomica Feltrina Rheticus for pointing out the citation that appears on page 154.*

# A FIREFLY BOOK

Published by Firefly Books Ltd. 1999
First published in Italian as *I Grandi Fenomeni Celesti: Le Comete, Le Stelle Cadenti, Le Eclissi* in 1997 by Arnoldo Mondadori Editore, S.p.A.

Copyright © Arnoldo Mondadori Editore, S.p.A.
English translation copyright © 1999 Arnoldo Mondadori Editore, S.p.A.

First Printing

**Library of Congress Cataloguing in Publication Data is available.**

**Canadian Cataloguing in Publication Data**
Vanin, Gabriele
  Cosmic phenomena : comets, meteor showers and eclipses

Translation of: I grandi fenomeni celesti: le comete, le stelle cadenti, le eclissi.
Includes bibliographical references and index.
ISBN 1-55209-423-5
1. Astronomy. 2. Comets. 3. Meteors. 4. Eclipses. I. Title.

QB500.V3613  1999    523    C99-930303-1

Published in Canada in 1999 by
Firefly Books Ltd.
3680 Victoria Park Avenue
Willowdale, Ontario
M2H 3K1

Published in the United States in 1999 by
Firefly Books (U.S.) Inc.
P.O. Box 1338, Ellicott Station
Buffalo, New York 14205

Art Director: Giorgio Seppi
Design, editorial production and drawings: Studio l'Atelier, Modena
English translation by Linda Eklund
English text edited by Christine Kulyk

Printed in Spain by Artes Gráficas Toledo, S.A.
D.L. TO: 848- 1999

# Contents

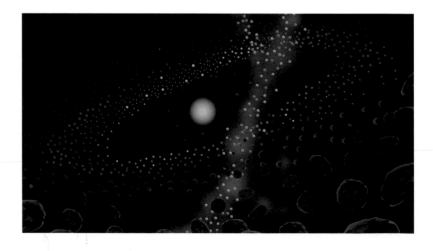

# Foreword

*The turn of the millennium is being heralded by a blaze of glory in earthly skies. Following two decades of disappointment, two Great Comets flashed through the northern firmament in 1996 and 1997. Then, 1998 brought the most impressive Leonid meteor shower of recent years, and 1999 again offers the possibility of a spectacular Leonid meteor storm. Finally, a total eclipse of the Sun in 1999 will be visible to millions of people as it slices across Europe and India from northwest to southeast—the first total solar eclipse to occur over these regions in 40 years.*

*Naked-eye astronomy offers many intensely interesting and wondrous celestial objects to our view. Consider the many fascinating faces of the Moon—from a thin, delicate crescent to the fat, full disk that sets the night aglow like a fairy tale; or the blazing light of splendid Venus that can even cast shadows on a dark night; or the multicolored glitter of Sirius, its light dispersed by passing through the layers of the Earth's atmosphere; or the intriguing and seductive configurations of constellations like Orion or star clusters like the Pleiades.*

*Three distinct categories of phenomena, however, surpass anything else in the sky for their majesty, sense of mystery and awesomeness. All these phenomena are easily visible to the naked eye. They can appear in more common forms—from the sudden flash of meteors on a tranquil night to a partial eclipse of the Sun or a total eclipse of the Moon. Or they can offer us the greatest spectacles of all, the grand cosmic phenomena with which this book is principally concerned—the appearance of a Great Comet, the fireworks of a spectacular meteor shower or the unsurpassed vision of a total eclipse of the Sun.*

*Yet many people may never witness even one of these spectacles in their entire lives—unless someone explains how to do it and why. And that is the scope of this book. It is written by one who has been lucky enough to witness wonders whose observation requires no optical instrument other than that precious one furnished by nature: the human eye.*

*Gabriele Vanin*

# The Great Comets

*Dazzling and restless bodies,
the Great Comets defy modern
astronomy's ability to predict
events. These celestial objects are
mysterious even now as, for the
most part, they stride on stage in a
completely unforeseeable way at
absolutely random intervals.*

*Observing these visitors, with
orbits so long that we will never
see them again in our lifetime, is a
great privilege which connects us
to the atavistic sensations of
humanity's first astronomers.*

# Great Comets of the Past

## What Is a Great Comet?

Due to their delicate beauty and the aura of mystery that has always surrounded them, comets are definitely among the most fascinating heavenly bodies. While ancient astronomers had developed complex models to predict the appearance of and alterations in the sky's other occupants, the enigmatic nature of comets prevailed. They showed up suddenly, stayed visible from a few days to a month, moved slowly among the constellations, then disappeared.

Where did they come from? What were they? Where did they go when they disappeared? The advent of a Great Comet—a comet whose tail reaches exceptional dimensions and whose head, or coma, shines as brightly as the brightest stars—was a powerful instigator of both curiosity and malaise, especially among the general public. In ancient times, the sudden appearance of such a monumental object had to be truly startling, rather as it is now. Even now that we know their nature, comets are still largely unpredictable objects capable of lighting up like headlights in the night and growing tails of incredible length and complexity or bursting like punctured balloons and disappointing everyone who waited for them with high hopes. This surely adds to their fascination.

A few dozen comets are discovered every year in modern times, but most can be seen only by telescope. On average (over the last three centuries for which complete and reliable statistics are available), a comet visible to

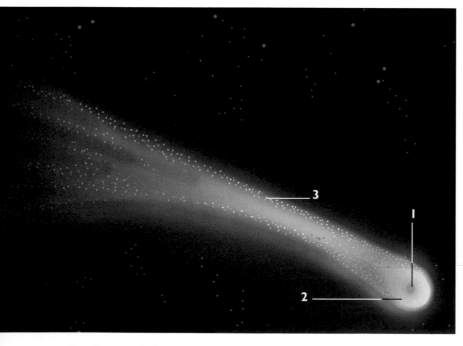

Preceding pages, the last Great Comet to appear before the recent Hyakutake and Hale-Bopp was Comet West (1976). Photo by Akira Fujii.

Above, the components of a comet: (1) the nucleus, or core; (2) the coma; (3) the tail.

Above right, how a Great Comet typically appears during maximum development. Here, the comet of 1843 is depicted over the skies of Paris. It was one of the most remarkable comets ever seen.

Left, a typical comet that reaches the brink of naked-eye visibility but never becomes a spectacular object. This was Comet Levy, which appeared in 1990, photographed by Eraldo Guidolin of the Galliera Veneta.

the naked eye has appeared once every two or three years; and only once in 10 years, more or less, does one appear that can be classified as a Great Comet.

It is obviously the latter that attracts the most attention from devotees and, above all, from the public at large. Unfortunately, Great Comets do not always appear on a schedule convenient for ordinary people. They are frequently visible only at dawn or immediately after sundown, and their period of peak visibility is often limited to just one week.

So how does a comet get to be a Great Comet? Contrary to what you might think, the absolute dimensions of the comet are not the only important factor. Its cometary activity also counts heavily—that is, the way it reacts to solar forces and how much of the surface of its nucleus is available to discharge volatile material. The comet's intrinsic luminosity depends

on this too. It must also contain an exceptional amount of water ice, whose sublimation feeds the activity of its nucleus when the comet approaches closest to the Sun (and when it appears the most luminous).

Typically, short-period comets (defined by convention as those with an orbital period under 200 years) are not very active, because they have lost most of their volatile surface material through numerous brushes with the Sun. Among the exceptions are Comet Halley and, to a lesser extent, two other comets that are perhaps younger or relatively larger than average. Generally speaking, only comets with long periods (thousands of years)—especially those that come from the distant confines of the solar system and are visiting the Sun's vicinity for the first time—have the potential to become Great Comets, because they possess an intact reserve of volatile elements. This is why astronomers

cannot predict the arrival of a Great Comet even today. Only after a promising comet is actually observed to be approaching can they anticipate what might happen in the succeeding few months.

Neither is it enough for a comet to be active. It must also pass close enough to the Sun and Earth—at a distance of, say, less than one astronomical unit (AU), the distance between the Sun and Earth (149.6 million kilometers). The closer the comet gets to the Sun, the more active it becomes; the closer it gets to Earth, the more luminous it appears (picture, for example, a 100-watt bulb seen from 100 meters away and then from only 10 meters). Many otherwise "ordinary" comets have become Great Comets thanks entirely to their favorable geometric positioning. For example, those of 302, 1106, 1402, 1680, 1843, 1882 and 1965 got close enough to the Sun to be visible in broad daylight. On the

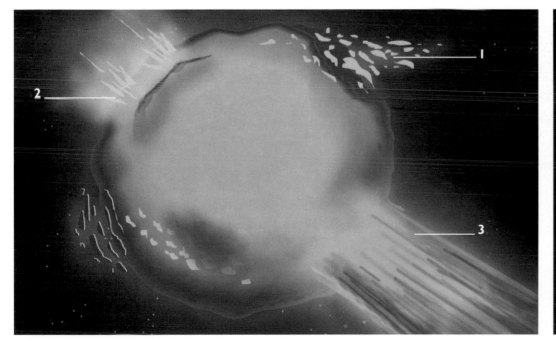

A comet may become more
or less active depending on
how its nucleus reacts to
solar forces.
1. Fragmentation of the crust
2. Surface explosions
3. Expulsion of gas and dust

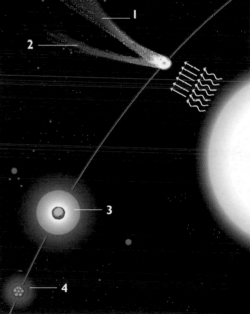

The closer a comet
approaches to the Sun, the
more active it becomes.
1. Dust tail
2. Plasma tail (also known
   as the gas, or ion, tail)
3. Coma
4. Nucleus

11

other hand, sometimes a close brush with Earth turns a comet into a Great Comet. This happened with the comets of 1132, 1471 and 1556, for example, as well as Comet Halley in 837 and 1066 and Comet Hyakutake in 1996.

Since a comet's impressiveness is determined far more by the length of its tail than by the brightness of its coma, however, the perspective from which the tail is viewed is critical.

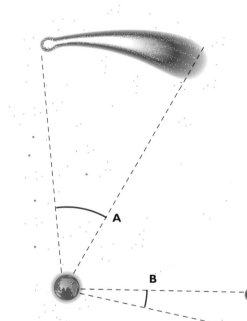

Thus even if a comet does not come near our planet but its tail is seen lengthwise, the comet can appear magnificent, like Comet Donati in 1858, for example.

Comets, as we shall see, usually exhibit two kinds of tails: a dust tail, which looks yellow-white because it simply reflects solar light, and

a gas tail, which is ionized by the Sun's ultraviolet radiation and hence is usually bluish in color. The gas tail shines due to a process of fluorescence, like the one that makes neon lights glow. Its colors are visible to the naked eye and by telescope, but only if the tail is extremely bright, which happens very rarely.

The gas tail, whose particles are flung out by the forceful pelting of the solar wind, usually appears straight and points directly away from the Sun. The dust tail, on the other hand, appears to curve, because the largest dust particles are not affected by solar radiation pressure and therefore tend to follow the nucleus in its orbital movement, while the lighter granules are subject to radiation pressure, so the comet's tail tends to spread out like a fan.

Of the two, the dust tail is usually a lot brighter, because it emits most of its light in the visible portion of the spectrum and our eye is

thus more sensitive to it. But the gas tail is generally longer, reaching its greatest length when the comet is at perihelion (the point where it is closest to the Sun), while the dust tail reaches its maximum development a few weeks later. As a result, a Great Comet is often more conspicuous after perihelion than before.

*Top, Comet Donati appears over Harvard Observatory.*

*Above, Halley's Comet photographed by Akira Fujii during its last visit. It is one of the few short-period comets that can become a Great Comet.*

*As shown in the diagram above, the apparent length of a comet's tail in the sky depends on its orientation with respect to the observer on Earth. Even though comet B actually has a longer tail than A, its apparent dimensions are smaller.*

*Facing page, by passing close to Earth, the recent Hyakutake became a Great Comet (shown here in a photo taken by the author with a 180mm lens).*

Even when a promising comet is detected, predictions about it are only approximate and must be taken with considerable caution. Too often, comet predictions have generated false hopes. A few cases must be considered among the most famous of this century. Comet Cunningham in 1940, Kohoutek in 1974 and Austin in 1990—long-period comets discovered many months prior to their closest approach—all seemed to promise a spectacular sight. Instead, as they approached the inner regions of the solar system, their brightness diminished so abruptly that they were barely visible to the naked eye. They must have been carrying very rich reserves of methane, ammonia and carbon dioxide, compounds that sublimate at a great distance from the Sun (explaining their initial intense luminosity), but precious little water.

## Tails and Comas in History

At this point, we will review the number and characteristics of Great Comets seen over the years and the sentiments they evoked, including what ancient people thought of these extraordinary objects. In times long past, comets were regarded in the European world as at-mospheric phenomena to which little importance was attached.

For many years, the beliefs of the great Aristotle (384–322 B.C.) held sway, and according to his doctrine, the heavens were immutable—nothing in them ought to change. Thus comets were not celestial bodies like the stars and planets but were instead hot, dry exhalations that the Sun's heat caused to erupt from underground. Rising toward the sphere of fire (placed above the sphere of air in Aristotelian cosmology), they caught fire due to friction produced by the upward movement. Rising even farther, the exhalation would reach the boundary between

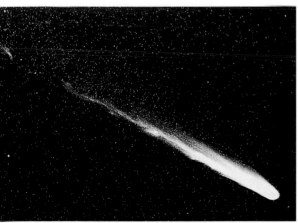

*Facing page, Comet Hale-Bopp with distinct dust and gas tails in a photo taken by the author.*

*The great Aristotle's notions influenced beliefs about comets through the end of the 17th century.*

*Above left, Kohoutek was supposed to become "the comet of the century" but ultimately proved to be a big dud. At its peak, it was only fourth magnitude, with a tail a few degrees long visible only by telescope and in photos. Photograph by Charles Kowal, taken with Palomar Observatory's 1.2-meter Schmidt camera.*

*Above, Austin—another "washout" comet—appeared in 1990, here photographed by the author.*

15

the sphere of fire and the first celestial sphere—that of the Moon—and would share the latter's movement around the Earth. The enormous authority accorded to the Stagirite philosopher in every sector of human knowledge assured the persistence of this notion until modern times. The first lists, or "cometographies," such as those of Poland's Johannes Hevelius (1668) or Flemish cometographer Stanislaus Lubienietz (1681), which reported a good number of sightings, display a striking degree of fragmentation, with many gaps and abundant tales of purely imaginary objects.

Fortunately, the astronomers of ancient China had fewer biases than the Europeans, and from 600 B.C. on, they carefully observed every bright comet visible to the naked eye, as well as novas, eclipses, planetary conjunctions, meteors and sunspots—in other words, all those phenomena that, with their mystery and inimitableness, festooned the destiny of the emperor and the empire with prophetic meaning. Ancient Chinese astronomy was, in fact, government astrology—organized into a powerful Astronomical Office where no fewer than a thousand people worked. The apparitions of comets, in particular, could have a happy or an unhappy meaning according to the circumstances of their arrival.

In Europe, too, comets were considered good or evil influences according to their location in the sky. And this was true across the ages. While ancient peoples tended to think of them positively, a negative interpretation prevailed during the Middle Ages. Comets were seen as the bearers of bad luck by Arabians and, across the ocean, by the Aztecs and Incas as well.

Top, the comet that, according to tradition, announced the arrival of Cortez, conquistador of Mexico, appears to Montezuma II. In reality, however, no Great Comet appeared in the 50 years preceding the arrival of the Spanish in the New World. In both the East and the West, the dates of rulers' deaths or of cometary appearances were freely falsified to demonstrate a connection between the two.

Above left, an 11th-century Chinese observation tower.

Above, one of the first lists of comets, or "cometographies," was the Theatrum cometicum of Stanislaus Lubienietz (frontispiece of the second section is shown).

## Historical Sources

The Chinese sky was divided into an unusually large number of constellations—almost 300—and the position of every observed phenomenon was identified according to the constellation in which it appeared. Thus comets were recorded with remarkable precision, without equal during the same period in Europe.

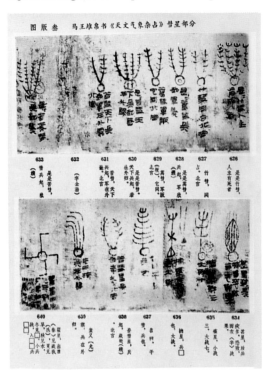

In his *Cometographie,* the French abbot Guy Alexander Pingre was able to use a catalog of comets observed in China and conveyed to Paris in 1759 by the Jesuit priest A. Gaubil. Published in 1783–84, this book was the first one written according to truly scientific criteria and is still valid today. Works this vast were never assembled again, but partial compilations have

surfaced that have also used precious, if somewhat less precise, Korean and Japanese observations dating to around the seventh century B.C. The most important and complete modern collections of cometary observations, widely consulted by us for this book, are those of Ho Peng Yoke (1962), Vsekhsvyatskij (1964), Hasegawa (1980), Kronk (1984), Hughes (1987) and Yeomans (1991).

From these and other offerings, we conclude that the total number of comets visible to the naked eye recorded from antiquity until now is about 1,250. Among them, just over 100 can be called Great Comets. These are listed in the table on page 29, which includes all comets mentioned in the above-noted sources that developed a tail at least 30 degrees long and a brightness of at least magnitude 2. Fainter comets to third magnitude, which did, however, develop undeniably spectacular tails, and

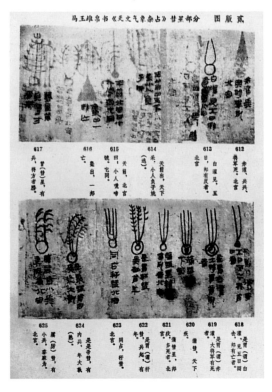

shorter-tailed comets that attained negative magnitudes were also taken into consideration. We adopted these criteria because this places a greater emphasis on the length of a comet's tail than on the brightness of its coma; what makes a comet great, in our view, is chiefly the size of the tail.

## Size and Brightness

The apparent length of a comet's tail (the length as it appears when seen from Earth by the naked eye, not the actual length) is expressed in degrees. One degree is 1/360 of the celestial sphere, 1/180 of the hemisphere from horizon to horizon and, of course, 1/90 of the distance between the zenith (the point directly above the observer) and the horizon. To illustrate this, the full Moon covers an angle of half a degree, while a hand span at the end of an outstretched arm covers an angle in the sky of about 20 degrees. Thus a 60-degree comet tail covers a third of the sky, while a 100-degree length extends over more than half the visible sky. While ancient observations in the European world were rather vague, Far Eastern records of comets were almost always accompanied by an indication of the length of their tails, expressed in a unit of measure called the *chhih,* or Chinese foot. It was equivalent to 1.5 degrees and was also adopted in Korea and Japan.

Chinese astronomers classified comets into various types according to their shapes. There were three fundamental categories: the *po-hsing,* or radiant stars (comets with no tail); the *hui-hsing,* or broom stars (comets with a tail); and the *chhang-hsing,* or long stars (comets

*Above, names and shapes of various types of comets, from* The Silk Book *of Ma Wang Tui, dating from the fourth century B.C. and found in a Chinese tomb from the second century B.C. Oldest known illustrations of* comets (courtesy of Richard Stephenson).

17

The ancient Chinese developed a constellation system that was completely different from our own, with three times as many star groups. This chart shows a sector of the sky with both Western and Chinese constellations

indicated for comparison.
• A North
• B East
• C West
A. Aries
B. Pisces
C. Pegasus
D. Triangulum

E. Andromeda
F. Lacerta
G. Cygnus
H. Perseus
J. Cassiopeia
1. Lou
2. Wai-Phing
3. Yu-Kêng

4. Khuei
5. Fên-Mu
6. Wei
7. Ying Shih
8. PI
9. Li-Kung
10. Chiu
11. Chhu

12. Thien Chiu
13. Thien-Ta-Chiang-
     Chün
14. Chün-Nen-Mên
15. Thêng-Shê
16. Wang-Liang

with an extremely long tail). These last two expressions were often used interchangeably, and sometimes all three terms were freely substituted one for another, so you may find a *po-hsing* cited with a tail dozens of degrees long and a *hui-hsing* with a very short appendage.

"Magnitude" is a number that expresses the apparent luminosity of any given star—or comet—using an inverted numeric scale: the bigger the magnitude number, the fainter the object. The brightest stars have the smallest magnitude numbers, those somewhat fainter have a slightly bigger magnitude number and on down the line to sixth-category stars, which are the faintest that can be seen by the naked eye. Luminosity changes from one magnitude to the next by a multiple of 2.5. For example, a first-magnitude star is 2.5 times as bright as one of second magnitude, 6.25 times brighter than a third-magnitude star, and so on. The scale also incorporates negative magnitude values, which express even greater brightness. Planets such as Venus and Jupiter have negative magnitudes, just as some comets occasionally achieve these exceptional measures of brilliance.

Nonetheless, defining the brightness of a celestial body that appears for the first time by comparing it with the brightness of a star or planet of known magnitude is a relatively recent custom. With few exceptions, it was practiced in neither the Far East nor Europe in the past. Thus the magnitude values reported in the table on page 29 are only rough indications. It is also important to remember, referring again to the table, that because of the geometric conditions of observation, only rarely did the maximum length of the tail coincide with the comet's maximum brightness.

## The Great Comets in Antiquity

Now let's take a brief look at those Great Comets listed in the table on page 29 that grew the longest tails and/or reached negative magnitudes or have assumed a particular role in the history of theories about these objects.

The first Great Comet definitely on record appeared in the constellation Scorpius in 147 B.C. and was observed by the Chinese. Its tail stretched halfway across the sky. It may have been the object that Seneca wrote about, describing it as large as the Sun, as red as fire and bright enough to dissipate shadows. The Roman philosopher was the only ancient theorist whose reasoning about comets was similar to modern thought and contrasted with the dominant Aristotelian line. He not only considered comets to be celestial bodies but also maintained that they had orbits similar to the planets. He went so far as to prophesy the advent of a man who would explain their nature.

This is all very remarkable if we remember that Seneca's views were confirmed only about two centuries ago. It is even more impressive that his prophecy, cast like a bridge across the millennia, would finally be realized in the person of Edmond Halley who, on the cusp of the 17th and 18th centuries, would embody the man invoked by the philosopher.

A comet with the same proportions as the comet of 147 B.C. appeared in 32 A.D. as well. According to the Chinese chronicles, "this was whitish blue in color, measured 60 to 70 [Chinese] feet long and 1 foot wide." To find records of other noteworthy apparitions, we must jump

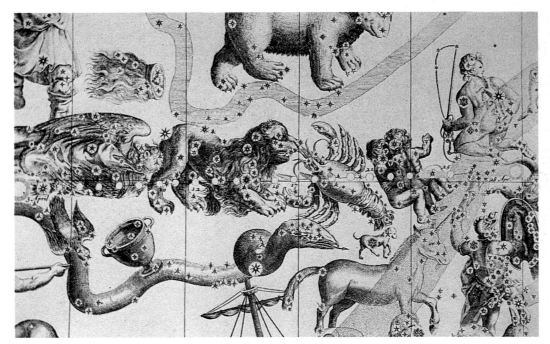

*The Roman philosopher Seneca held extraordinarily precocious beliefs on the nature of comets.*

*Course of Halley's Comet across the sky in 837 during the closest of its 30 historically recorded journeys past the Earth, as illustrated in the Theatrum cometicum of Stanislaus Lubienietz.*

19

stars of Chung-Tai [these three groups are a part of our constellation Ursa Major]." The comet was observed in Rome, as well, and remained visible for 100 days altogether.

## Apparitions in the Middle Ages

In 837, the world stood dumbfounded at the sight of an extremely bright comet whose tail crossed two-thirds of the celestial dome and whose head shone brighter than Jupiter. It moved so fast that it traversed the whole sky in just five days. It was Halley's Comet, in the closest of all its historical transits (less than six million kilometers from Earth). It was seen for only 37 days, because by then, it was moving away after passing perihelion. For a few days, its tail divided into two branches that were highly

ahead more than two centuries. Two enormous comets appeared in 178 and 191 A.D. The first was visible for 80 days, exhibiting a tail that stretched 70 to 90 degrees, while the second literally seemed to fill the entire sky. Two especially striking comets appeared in 252 and 253–54, with tails 80 degrees long. The second stayed visible to the naked eye for at least 190 days. In 287 in China, a comet was observed with a tail that, according to the chronicles, "measured hundreds of feet" (!?).

In 374, we have the first truly spectacular return of Halley's Comet. On that visit, thanks to its extreme proximity to Earth, it developed a 100-degree tail and exceeded the magnitude of Jupiter (–3)—the brightest body in the sky after the Sun, Moon and Venus. In 390, a comet observed in China, Korea and Rome while it was moving from Gemini toward Ursa Major displayed a 150-degree tail and reached a magnitude of –1. In 418, a comet again reached these incredible dimensions when—as several Chinese chronicles put it—it was situated "west of Thai-Wei [corresponding to the western part of Virgo] . . . and its rays spread more than 100 feet, passing through Pei-Tou, Tzu-Wei and the

*The parallax method, used on comets for the first time by Georg von Peurbach, allows calculation of a celestial object's distance by observing it from two positions sufficiently far apart.*

*Above, Halley's Comet of 1066 illustrated in the Bayeux tapestry.*

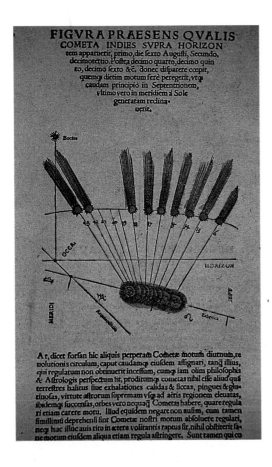

when it is visible in the morning and east when it is visible in the evening, but there has never been one that pointed in the four directions and passed through as many constellations as this comet."

In 891, another object of monstrous proportions appeared in Ursa Major less than two years after Chinese astronomers reported a comet with an amazing 300-degree tail. Perhaps there is an error of transcription in the journal, because it is hard to imagine not only a comet that covers the whole firmament but a comet whose tail continues to be visible above the horizon for quite a while after its head has

already set! The same might be said of the comet of 287 cited above.

In the favored year 1402, two brilliant comets appeared, the first visible from February through April and the second from June through September. Because both were visible in daylight, they had to have reached a minimum magnitude of –5. The first, in particular, remained visible for eight consecutive days. As for tail length, the first was rather short, but the second was described as "immense." Since the latter was seen only in Europe, its arrival probably coincided with the monsoon season in the Far East.

resplendent, and at a certain point, the coma dipped below the horizon for midnorthern observers, with only the tail poking above. The tail was extremely long, like a flame leaping up from the ground. An ancient Chinese journal, the *Wên Hsien Thung Khao* ("Historical Review of Public Affairs," compiled by Ma Tuan Lin in 1254), was usually very reserved in tone when recounting apparitions of even the most exceptional celestial objects. With a sense of awe, it reported: "It is normal for a comet to point west

*Top, having observed various comets—among them the Great Comet of 1533— Apianus, along with Regiomontanus, recognized the antisolar nature of comets' tails. The concept is very well depicted here for Halley's Comet during its 1531 visit in drawings taken from Apianus's Astronomicum Caesareum (Ingolstadt, 1568).*

*Above, in his Adoration of the Magi, at Padua's Scrovegni Chapel (1304–5), Giotto painted a very realistic comet and may have taken his inspiration, at least in part, from the apparition of the Great Comet of 1299.*

## The Modern Epoch

In 1456, Austria's Georg von Peurbach was the first to subject Aristotle's beliefs to experimental proof by trying to measure the parallax, and hence the distance, of a comet which appeared that year. The method is rather simple conceptually. Von Peurbach and one of his colleagues observed the comet from locations 10 kilometers apart. If it were as close to Earth as Aristotle proposed, then the comet would appear to occupy a different position for each observer with respect to the backdrop of fixed stars. On the other hand, if it were farther away—say, beyond the Moon—the two observers would not notice a change in position or they would notice more modest changes according to the accuracy of the measurements taken and the instruments used. The amount of change in position could also yield, indirectly, the respective distances of the comet and the Moon from Earth. It seems, however, that von Peurbach was more interested in measuring the altitude of the sphere of fire (where, according to Aristotle's cosmological model, the comets set themselves on fire) than in measuring the comet's distance. The comet, in this case, was none other than Comet Halley in one of its historically documented passes.

On Christmas Day 1471, a comet was noticed in Europe and followed in Korea until February 17, 1472. It presented a 50-degree tail, and due to its great proximity to the Earth (just over 10 million kilometers on January 23), it became visible during the day. It was the last of five comets observed by Paolo dal Pozzo Toscanelli,

*The Comet of 1577, observed by Tycho Brahe, was the first for which precise parallax measurements were computed (depicted here in a 16th-century Turkish manuscript, courtesy of Richard Stephenson).*

the celebrated Italian geographer whose beliefs about the width of the Atlantic Ocean encouraged the great voyage of Columbus. With the first comet he observed in 1433, Toscanelli introduced to the Western world the practice of making precise measurements of comets. He measured both the position of the head among the stars and the direction and length of the tail.

Johann Müller (called Regiomontanus), a student of von Peurbach, attempted to calculate the parallax and distance of the comet of 1472. He found a value of nine Earth radii. This damaged but did not dethrone Aristotle, because it meant the comet was still positioned below the first celestial sphere, that of the Moon.

The comet of 1533 was very bright, reaching a magnitude of –2, although it did not display a particularly long tail. It was seen by Copernicus, Girolamo Fracastoro of Verona and Germany's Peter Bienewitz (called Apianus). From their observations, Apianus and Fracastoro noticed that comets' tails always pointed away from the Sun, and they therefore deduced that the nature of these objects must have some kind of relationship with the Sun. This fact had already been recorded in China as much as nine centuries earlier, but in Europe, it acquired an unusual importance in light of the conflict it created with the reigning Aristotelian concept.

In 1556, another comet passed very close to the Earth—12 million kilometers—reaching a magnitude of –2. Even though its tail was rather short, it was called "frightening" and "miraculous."

On November 1, 1577, the comet that would change Europe's notions about comets was seen from Peru. It was also one of the most dazzling ever. One Japanese source said: "It was as bright as the Moon." It was certainly brighter than Venus. It was visible even through clouds, just like the Moon, and it displayed a tail as long as 75 degrees. Late on the afternoon of November 13, Denmark's Tycho Brahe, then 31 years old and not yet the greatest astronomical observer of all time, was going fishing near his observatory at Uraniborg when he saw an extremely bright object which could have been Venus, but he knew that the beautiful planet was a morning "star" during that period. After sunset, the "star" revealed itself for what it was—an extremely bright comet with a tail more than 20 degrees long.

It was seen all across Europe and produced an unprecedented theoretical debate. Only five years earlier, Tycho, the "terrible Dane," had observed a new "star" in Cassiopeia that, from parallax measurements, clearly seemed to belong to the celestial spheres. The Aristotelian universe was vacillating; it was no longer true that the heavens were immutable. From precise parallax measurements taken on the comet, Tycho calculated that it was at least six times farther away than the Moon. Though less accurate, parallax measurements were also obtained by other famous astronomers of the time, including Mästlin, Kepler's teacher, and they all

Above, the comet of 1556 over Constantinople in a color etching by Herman Gall.

Right, the writings of Denmark's Tycho Brahe about planet Mars, the supernova of 1572 and the Great Comet of 1577 constitute perhaps the greatest observational contribution to astronomy by any single observer.

Left, Edmond Halley pictured in 1721 when, at age 65, he had recently become the British Astronomer Royal.

23

pointed in the same direction. They contradicted the Aristotelian belief in comets as sublunar phenomena and helped develop their status as heavenly bodies.

In 1618, three comets appeared, the brightest of which was observed from November 1618 until January 1619, was visible even during the day and had a tail of gigantic proportions. Galileo observed all three but used them rather infelicitously in a sterile polemic with the Jesuit priest Orazio Grassi in which the Pisan sustained erroneous and outdated beliefs. With that, the cometary debate reached a momentary stalemate. Meanwhile, another comet with a 100-degree tail was observed in 1619.

In 1664 and 1665, two comets appeared that were neither especially bright nor particularly conspicuous but were destined to stimulate the publication of a remarkable number of scientific works. They were observed by the most acclaimed astronomers of the age: Hevelius, Huygens, Cassini, Borelli and Hooke. While Hevelius and Huygens maintained that comets were objects of an ephemeral nature endowed with a uniform straight-line motion, Hooke believed them to be permanent celestial bodies moving in straight lines. Cassini and Borelli were the most "modern" in affirming that the comets followed a closed orbit (although Cassini maintained that the comet of 1664 rotated not around the Sun but around the star Sirius).

The comet of 1680 was bright enough to be visible in broad daylight 2 degrees from the Sun. A memorable record of this appearance can be found in the handwritten diary of Bernardino Pagani, preserved at the Civic Library in Belluno: "I went out into the countryside with many other people to see a star that others had already observed for several days. I beheld it as a truly monstrous and fearsome thing, and anything but little. To our eyes, it was a giant and extremely long whorling ribbon that proved truly horrifying."

This comet was the first to be discovered by telescope (by Gottfried Kirch) and the first for which a parabolic orbit was calculated (by Georg Dörffel). After it had been visible by daylight for two days, British Astronomer Royal John Flamsteed saw only its tail in the evening sky, while the coma reappeared two days later as it moved farther from the Sun. The tail reached 80 degrees in length. The last to observe it, by telescope, was Isaac Newton, a year after its discovery. He was initially convinced that comets moved in straight lines but later changed his mind, thanks in part to discussions he had with Edmond Halley. The latter, seeking to convince Newton to publish his work on the principles of universal gravitation, took particular interest in the comet that appeared in 1682. It was not very bright, but by studying its orbit intensely, along with the orbits of 23 other comets which had appeared in the preceding three centuries, Halley realized that its orbital elements were very similar to those of the comets of 1531 and 1607 and that it was therefore the same comet traveling a closed elliptical orbit which lasted about 76 years.

It was a fundamental discovery that finally ratified the celestial status of comets. The British scientist risked predicting the return of "his" comet in 1759, a prediction that came true right on schedule, 14 years after his death. Thus 17 centuries after Seneca, Edmond Halley emerged as the man who would verify Seneca's theory.

## The 16th and 17th Centuries

The most spectacular comet in the 1700s was Comet de Cheseaux, named for the Swiss astronomer who studied it at length but did not discover it. This comet was an absolutely unique and stunning object, an apparition without equal in the history of observations. By February 1, it was at its most brilliant (magnitude –1.4); by February 18, it shone like Venus and displayed two tails, one 7 degrees long and the other 25 degrees. But on March 7 and 8, when its coma sank below the horizon, de Cheseaux was dazzled by the phantasmagoric

*Right, anonymous watercolor depicting the comet of 1680 over the Low Countries calls to mind Flamsteed's awestruck observation cited in the text above.*

*Far right, Comet Donati portrayed over Notre Dame cathedral in Paris.*

spectacle of a multiple tail divided into six branches that rose like a fountain gushing from the horizon. The comet also became visible in broad daylight.

On August 8, 1769, the great Charles Messier—the "comet ferret," as King Louis XV liked to call him—detected the only one of his 13 discoveries that would become a Great Comet. It reached zero magnitude, grew a tail almost 100 degrees long and could be seen by the naked eye for a minimum of 100 days.

In 1811, a comet appeared whose influence was said to cause an extraordinary grape harvest. For years, the wines bottled that year were considered exquisite, and even today, enologists and cognoscenti consider 1811 "the year of the comet." The comet did not pass particularly close to our planet or the Sun (its perihelion lay outside the Earth's orbit). With its high intrinsic luminosity, however (probably produced by a nucleus of unprecedented dimensions), it still achieved zero magnitude and stayed visible to the naked eye for all of 260 days, a record exceeded only recently by Comet Hale-Bopp. Its tail was split into two forks, one evidently made of dust and the other of gas, each of which was no longer than 25 degrees. Due to the great distance between the Earth and the comet, however, this translated to a tail that was over 150 million kilometers long. The dust tail was also very wide, at least 7 degrees.

In late February 1843, an extremely bright comet was seen in New England during the day at 1 degree from the Sun. A close brush to just 120,000 kilometers from the Sun's surface produced an apparent luminosity of at least –17, equal to 50 times the brightness of the full Moon. In early March, it was visible in the evening with a very dense, bright tail almost 70

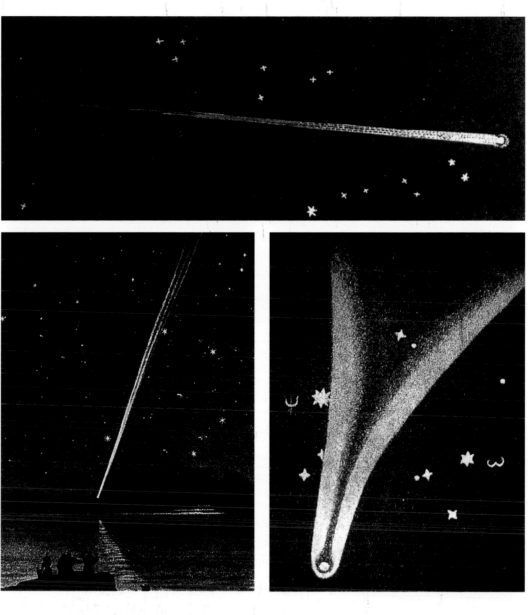

degrees long and a head that still shone like Jupiter. The true length of the tail was 320 million kilometers. The comet appeared in the same year that a certain William Miller had designated as the end of the Earth. The coincidence reawakened ancestral fears that disappeared together with Miller's followers when

nothing happened on April 3, 1843, the date specified for the apocalypse.

In 1858, the most beautiful comet within human memory appeared. Discovered by Florentine astronomer Giovanni Battista Donati, it had a wide, curved tail of dust and one of gas, which were at first united but later broke into

*Top, depicted in a contemporaneous drawing, the tail of the comet of 1861—at least 120 degrees—was the longest ever registered in the modern era.*

*Above left, the Great Comet of 1843 drawn by Charles Piazzi Smyth at the Cape of Good Hope. Its reflection in the sea is especially evocative.*

*Above, the comet of 1811 is illustrated in this drawing by Austrian astronomer Franz Xaver von Zach.*

two very thin prongs. The tail extended 60 degrees at its longest and sped through a series of rapid alterations (calling to mind the changing appearance of the recent Comet Hyakutake's tail). In one day, it grew from 35 to 50 degrees long, and in just three days, it shriveled from 45 to 15 degrees. The coma also put on a show: Observed through a telescope, it exhibited a structure of jets forming a rotating pinwheel, changing continually from night to night. Comet Donati was the first comet to be photographed—not by an astronomer but by an English photographer who caught it with a simple portrait camera. Only the brightest part of the coma and tail were visible in the picture, however.

In 1861, a comet was recorded at less than 20 million kilometers from Earth. This produced an enormous apparent elongation of its tail, which stretched from the star Beta Aurigae all the way to the constellation Ophiuchus—at least 118 degrees—the longest ever seen in modern times, although its true length was far

more modest, at about 50 million kilometers. This comet was first seen in the southern hemisphere, but news of its discovery did not reach Europe before the comet had appeared by surprise in the morning sky on the last day of June. It was so magnificent that one of the first people to see it, not knowing what it might be, thought it was the rising Moon.

The following year, there was a favorable passage of Comet Swift-Tuttle, the only periodic comet besides Halley that can aspire to the role of Great Comet. The Perseid meteors, the famous August 11 shower of "falling stars," come from the dust of this comet. It was not an especially brilliant comet, but it developed a 30-degree tail by the end of August 1862. During its recent reappearance in 1992 (the comet has a period of 130 years), however, it was barely visible to the naked eye and its 2-degree tail could be seen only by telescope.

In 1882, the comet with the longest tail (in real terms) in history appeared. Remembered as the Great September Comet, it was discov-

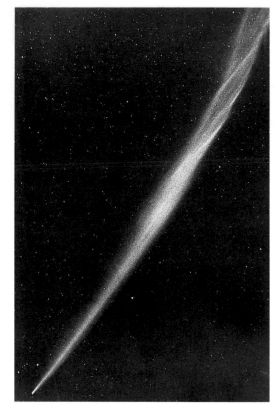

ered in New Zealand the second day of that month. Like the comet of 1843 mentioned above, this was a sun grazer. On September 16, it was visible by telescope at the Cape of Good Hope all day long. The next day, when it was supposed to pass perihelion, it was seen to approach the brink of the Sun and . . . disappear, the first unassailable sign of the extremely small size of cometary nuclei and the extreme rarefaction of their comas and tails. It traveled to just 430,000 kilometers from the Sun's surface. It was visible to the naked eye in daylight until September 20, becoming 10 times brighter every day and finally reaching a peak magnitude of –18. Then it diminished in brightness but began to produce a stupendous,

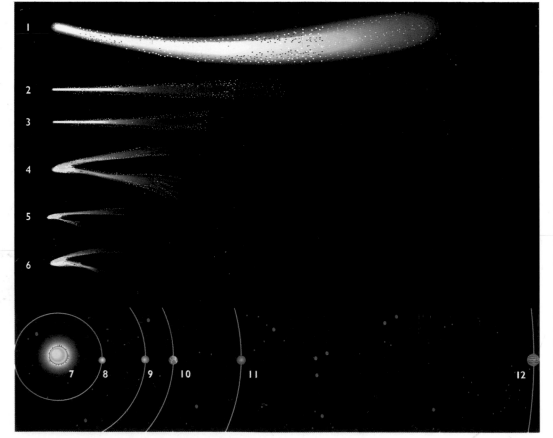

In the illustration at left, some of the longest comet tails are compared with the radii of planetary orbits.
1. 1882, 600 million km
2. 1843, 320 million km
3. 1680, 240 million km
4. 1811, 176 million km

5. 1910 (Daylight), 110 million km
6. Hale-Bopp, 100 million km
7. Sun
8. Mercury
9. Venus
10. Earth
11. Mars
12. Jupiter

broad, curving tail. On November 21, it reached its maximum length of about 600 million kilometers; at this point, the comet's head was at the orbit of Mars and the end of its tail was at Jupiter's. (An unfavorable perspective and the great distance, however, reduced its apparent length to less than 30 degrees.) Its nucleus also shattered into four fragments.

## The Most Recent Great Comets

In 1910, there were two Great Comets. The first, universally known as the Great Daylight Comet, appeared in January in the southern hemisphere, and many people confused it with Halley's Comet, whose return was expected imminently. It had already been recovered telescopically in the preceding months, and everyone knew that it would become visible to the naked eye. The Great Daylight Comet was seen by telescope from Capetown on January 17 just 3 degrees from the Sun and, after January 20, was also visible, with a coma as bright as a first-magnitude star, from midnorthern latitudes. On the night of January 29, its dust tail put on a spectacular show, splaying out like a giant fan. Halley's Comet was much less spectacular. Although its tail reached 100 degrees in May thanks to a favorable point of view, its head never got any brighter than zero magnitude. On the other hand, that apparition was 100 times more favorable than the most recent one, in 1986. Due to unfavorable geometric circumstances that year, the most famous of comets never exceeded magnitude 2 and had a very faint tail visible to the naked eye but never longer than 10 degrees.

There was another daylight comet in 1927 called Skjellerup-Maristany, which appeared 5 degrees from the Sun on December 18, while its tail reached 40 degrees toward the end of the year. Unfortunately, it was visible only from the southern hemisphere.

In 1948, in Nairobi, Kenya, just as the last sliver of Sun disappeared during the November 1 solar eclipse, there was a ghostly apparition. A brilliant comet cut into the plumes of the solar corona. Its coma lay just 1.5 degrees from the Sun's rim, and it had a long, curving tail. The object was first seen again after the eclipse on November 4 by a Pan Am pilot flying over Jamaica. In the following days, its brightness was estimated at first magnitude, with a tail 30 degrees long. It stayed visible to the naked eye until about the middle of December.

Nineteen fifty-seven has come to be remembered as the year of the comet with an anti-tail. By the end of April, Comet Arend-Roland, in addition to exhibiting a splendid dust tail nearly 30 degrees long, grew another extraordinary tail as long as 15 degrees that curved toward the Sun. This phenomenon is produced when the Earth crosses the orbital plane of an active comet. Dust from the comet that lies in this plane is rendered visible as it curves beyond the comet and seems to point toward the Sun. But the whole effect is merely an illusion of perspective. Discovered eight months before it reached perihelion, Arend-Roland was the second comet this century to cause great expectations. Fortunately, unlike Comet Cunningham 17 years earlier, everything went well this time.

The comet of 1965 was baptized the "comet of the century." Discovered by two Japanese stargazers on September 18, Comet Ikeya-Seki was already visible to the naked eye in early October. On October 13, it was third magnitude, and its luminosity grew by half a magnitude each day until October 21, when it was expected to pass perihelion at only 450,000 kilometers from the Sun's surface. In the last 24 hours before perihelion, the comet was easily seen by the naked eye in full daylight (by simply blocking the Sun with the hands), and it attained approximately the brightness of Venus. In Arizona, D. Meisel was even able to see a 2-degree tail with the unaided eye. In California, C. Capen saw a 4-degree tail by telescope. In Japan, where the comet was supposed to reach perihelion at noon, local time, the coma was estimated at magnitude –17 and was seen less than 1 degree from the Sun's rim. From the morning of October 25, having left the Sun's

*Facing page, top, Comet Ikeya-Seki photographed by Eugene Harlan on October 29, 1965, at Lick Observatory.*

*Above, the synchronous bands of Comet West were captured in this beautiful image by Japanese astrophotographer Akira Fujii.*

embrace, Ikeya-Seki unfurled a bright tail that extended more than 20 degrees (at least 45 degrees, according to some observers). It was more lustrous than the Milky Way's star clouds in Sagittarius—so brilliant that you could see it through thick fog cover—narrow and with a helical internal structure. Around the first week of November, its nucleus broke into two or three pieces.

Comet Bennett, which appeared in 1970, flourished a dust tail that was extremely bright, although not very long (20 to 25 degrees), while its ion tail was much fainter and never exceeded a dozen degrees. Even the coma was rather bright; it reached a negative magnitude and, telescopically, exhibited some peculiar and delightful spiral jets very similar to those observed on Comet Donati. Already visible to the naked eye in the early days of February, it became a conspicuous object after mid-March. It was one of the few Great Comets visible for the better part of the night and became circumpolar, visible throughout the night toward the end of April. It remained visible to the naked eye until the end of May.

The last truly Great Comet before the advent of Hyakutake was Comet West in 1976. It was at magnitude 15 when European Southern Observatory astronomer Richard West discovered it on November 5, 1975. Calculations of its orbit suggested a very favorable apparition in the months preceding perihelion; but after the disappointment of Kohoutek two years earlier, no one was willing to risk a prediction. Nonetheless, Donald Yeomans of the Jet Propulsion Laboratory gambled on a coma magnitude of 0.8 for the morning of March 1, 1976. He was badly mistaken but, fortunately, erred on the low side. Bortle saw Comet West unaided on the day of perihelion, February 25, just before sunset, 7 degrees from the Sun.

Comet West sprang spectacularly from the horizon in early March, with a head that was still magnitude −1. It marshaled two well-defined and extremely bright tails, with the different colorations of dust and gas tails easily visible to the naked eye. While the ion tail took the form of a short 5-to-10-degree spike pointing upward, the dust tail opened out into a huge fan with a 30-degree spread. The fan displayed some peculiar structures called synchronous bands that gave the tail the look of

window blinds. Gil Wood may have made the most impressive observation of Comet West. He was climbing Mount Pinos on March 7 and first saw the comet at 2,700 meters: "I was just getting through the last curves near the summit when I looked out the car window, and there it was—a fantastic fountain of light that peeped through the tops of the trees. The head of the comet was too low to be seen above the embankment around the parking lot at the summit, but the tail opened like a fan superposed on the Milky Way in Cygnus." The show continued through March. On March 15, the tail was still 15 degrees long, with a third-magnitude coma. West's nucleus split into four parts (a further clue to the extreme friability of a comet's core), some of which were observed until April.

For 20 years after Comet West, no other noteworthy comet appeared. Statistically, this is an unusually long period of drought. In effect, a whole generation of skywatchers was deprived of seeing a Great Comet. Happily, the arrival of comets Hyakutake and Hale-Bopp served to fill this void rather grandly and in the absolutely unpredictable manner that comets are so good at.

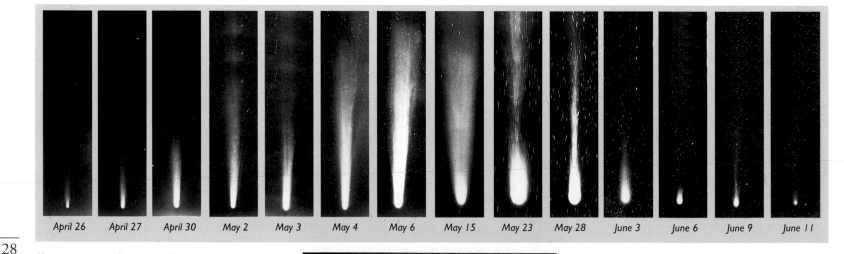

| April 26 | April 27 | April 30 | May 2 | May 3 | May 4 | May 6 | May 15 | May 23 | May 28 | June 3 | June 6 | June 9 | June 11 |

*Above, progression of Comet Halley's tail during its 1910 apparition. (Hale Observatories)*

*Right, Comet Bennett in a beautiful panoramic image taken by the Geneva Observatory on March 26, 1970.*

**GREAT COMETS** from 238 B.C. to 1402 A.D.

| Date | Tail | Magnitude | # of Days Visible | Peri-helion | Perigee | Name |
|---|---|---|---|---|---|---|
| 238 B.C. | * | | 80 | | | |
| 214 B.C. | * | | 80 | | | |
| 147 B.C. | 90° | ∞ | 32 | 0.43 | 0.15 | |
| 134 B.C. | * | | 70 | | | |
| 32 B.C. | 90° | | | | | |
| 39 A.D. | 45° | | 49 | | | |
| 65 | 36° | | 56 | | | |
| 101 | 45° | | 10 | | | |
| 133 | 75° | | | | | |
| 178 | 80° | 2 | 80 | 0.50 | | |
| 191 | 150° | | | | | |
| 240 | 45° | 1–2 | 40 | 0.37 | 1.00 | |
| 252 | 80° | | 20 | | | |
| 253 | 75° | | 190 | | | |
| 254 | # | | | | | |
| 287 | § | | 10 | | | |
| 302 | | @ | | | | |
| 374 | 100° | –3 | 30 | 0.58 | 0.09 | Halley |
| 390 | 150° | –1 | 10 | 0.92 | 0.10 | |
| 400 | 45° | 0 | 30 | 0.21 | 0.08 | |
| 416 | $ | | 80 | | | |
| 418 | 150° | | 100 | | | |
| 420 | * | | | | | |
| Feb. 423 | 45° | | 20 | | | |
| Dec. 423 | 60° | 0 | 10 | | | |
| 451 | | –3 | 60 | 0.58 | 0.49 | Halley |
| 454 | 30° | | | | | |
| 467 | * | | | | | |
| 607 | * | –2 | 30 | 0.58 | 0.09 | Halley |
| 608 | 40° | | | | | |
| 676 | 45° | | 58 | | | |
| 760 | # | 0 | 50 | 0.58 | 0.41 | Halley |
| 770 | 75° | 1–2 | 60 | 0.58 | 0.30 | |
| 837 | 120° | –3 | 46 | 0.58 | 0.03 | Halley |
| 838 | 52° | | 48 | | | |
| 852 | 75° | | | | | |
| 891 | 150° | | 50 | | | |
| 893 | 300° | | 37 | | | |
| 905 | 60° | 0 | 25 | 0.20 | 0.21 | |
| 907 | 45° | | 8 | | | |
| 939 | æ | ∞ | 8 | | | |
| 975 | 60° | | 83 | | | |
| 989 | $ | 0 | 30 | 0.58 | 0.39 | Halley |
| 1018 | 60° | | | | 0.38 | |
| 1019 | 45° | | 37 | | | |
| Sept. 1041 | 45° | | 20 | | | |
| Nov. 1041 | 45° | | 10 | | | |
| 1056 | 30° | | | | | |
| 1066 | 25° | –4 | 60 | 0.58 | 0.10 | Halley |
| 1097 | 50° | | 33 | | 0.52 | |
| 1106 | 150° | –15 | 40 | 0.008 | 1.2 | |
| 1127 | & | | 29 | | | |
| 1132 | 45° | –1 | 20 | 0.74 | 0.04 | |
| 1145 | 70° | –3 | 81 | 0.58 | 0.27 | Halley |
| 1146 | 150° | | | | | |
| 1222 | 45° | 1–2 | 35 | 0.58 | 0.31 | Halley |
| 1223 | 30° | | | | | |
| 1232 | 60° | | 58 | | | |
| 1264 | 150° | –2 | 80 | 0.82 | 0.18 | |
| 1299 | 30° | | | | | |
| Mar. 1362 | 30° | | 33 | | 0.44 | |
| Apr. 1362 | # | | 40 | | | |
| 1385 | 30° | | | | 0.74 | |
| Feb. 1402 | 15° | –5 | 70 | 0.38 | 0.71 | |

**GREAT COMETS** from 1402 A.D. to 1997 A.D.

| Date | Tail | Magnitude | # of Days Visible | Peri-helion | Perigee | Name |
|---|---|---|---|---|---|---|
| June 1402 | Δ | @ | 90 | | | |
| 1439 | 75° | | | | 0.31 | |
| 1456 | 60° | 0 | 42 | 0.58 | 0.45 | Halley |
| 1465 | 45° | | 90 | | | |
| 1468 | 45° | 1–2 | 81 | 0.85 | 0.67 | |
| 1472 | 50° | @ | 58 | 0.49 | 0.07 | Regiomontanus |
| 1533 | 15° | –2 | 81 | 0.25 | 0.42 | Apianus |
| 1538 | 45° | | | | 0.94 | |
| 1556 | 7° | –2 | 73 | 0.49 | 0.08 | Heller |
| 1577 | 75° | –7 | 79 | 0.18 | 0.63 | Brahe |
| 1578 | 80° | | | | | |
| 1580 | 10° | –5 | 106 | | 0.23 | |
| 1582 | 150° | | 20 | | 0.83 | Brahe |
| 1585 | 2° | –4 | | 1.095 | 0.14 | |
| 1587 | 50° | | 90 | | | |
| 1618 | 104° | @ | 67 | 0.40 | 0.36 | |
| 1619 | 100° | | 30 | | | |
| 1664 | 37° | 1 | 75 | 1.03 | 0.17 | Hevelius |
| 1665 | 30° | –4 | 24 | 0.11 | 0.57 | Hevelius |
| 1668 | 40° | 2 | 27 | 0.07 | 0.80 | Gottignies |
| 1680 | 80° | @ | 80 | 0.01 | 0.42 | Kirch |
| 1682 | 30° | 0 | 40 | 0.58 | 0.42 | Halley |
| 1686 | 35° | 1 | 30 | 0.34 | 0.33 | |
| 1689 | 68° | 3.5 | | 0.06 | 0.75 | |
| 1695 | 40° | | 22 | 0.01 | | Jacob |
| 1702 | 43° | 0 | 10 | 0.01 | | |
| 1744 | 90° | –7 | 45 | 0.22 | 0.83 | de Cheseaux |
| 1769 | 98° | 0 | 100 | 0.12 | 0.32 | Messier |
| 1811 | 25° | 0 | 260 | 1.04 | 1.22 | Flaugergues |
| 1835 | 30° | 1 | 148 | 0.59 | 0.19 | Halley |
| 1843 | 68° | –17 | 48 | 0.006 | 0.84 | Great March Comet |
| 1858 | 60° | 0.5 | 80 | 0.58 | 0.54 | Donati |
| 1861 | 118° | 0 | 90 | 0.82 | 0.13 | Tebbutt |
| 1862 | 30° | 2 | 60 | 0.96 | 0.34 | Swift-Tuttle |
| 1864 | 40° | 2 | 30 | 0.91 | 0.10 | Tempel |
| 1874 | 66° | 0 | 50 | 0.68 | 0.29 | Coggia |
| 1880 | 50° | 3 | | 0.005 | 0.69 | Great Southern Comet |
| Mar. 1882 | 45° | –5 | 60 | 0.06 | | Wells |
| Sept.1882 | 30° | –18 | 135 | 0.008 | 0.99 | Great September Comet |
| 1887 | 52° | 1.5 | | 0.005 | 0.60 | Thome |
| 1901 | 30° | –1.5 | 38 | 0.24 | 0.83 | Viscara |
| Jan. 1910 | 40° | –4 | 17 | 0.13 | 0.86 | Great January (Daylight) Comet |
| Apr. 1910 | 100° | 0 | 80 | 0.59 | 0.15 | Halley |
| 1927 | 40° | –6 | 32 | 0.18 | 0.75 | Skjellerup-Maristany |
| 1947 | 30° | 0 | | 0.11 | 0.85 | Southern Comet |
| 1948 | 20° | –3 | 45 | 0.14 | 0.55 | Eclipse Comet |
| 1957 | 45° | 1 | 45 | 0.32 | 1.2 | Arend-Roland |
| 1962 | 20° | –2.5 | 57 | 0.03 | | Seki-Lines |
| 1965 | 45° | –17 | 30 | 0.008 | 0.91 | Ikeya-Seki |
| 1970 | 25° | –0.3 | 80 | 0.54 | 0.69 | Bennett |
| 1976 | 30° | –6 | 55 | 0.20 | 0.79 | West |
| 1996 | 50° | 0 | 40 | 0.23 | 0.10 | Hyakutake |
| 1997 | 15° | –1 | 300 | 0.91 | 1.31 | Hale-Bopp |

*Date of apparition is given in the first column; maximum tail length and magnitude in the second and third. Column 4 gives the interval of naked-eye visibility; columns 5 and 6, the minimum distances from the Sun and Earth in astronomical units; column 7, the comet's name (often that of its discoverer).*

$ *long*
æ *very long*
* *across the sky*

# *dozens of degrees long*
§ *hundreds of Chinese "feet" long*
Δ *immense*
& *took up the whole sky*
@ *visible in daylight*
∞ *lit up shadows*

# The Origin and Nature of Comets

## The Birth of the Solar System

By now, you may be wondering: Where do comets come from? And what are they? To respond adequately, we have to begin with the birth of our solar system about 4.5 billion years ago. This took place 25,000 light-years from the center of our galaxy, where an interstellar nebula (a region rich in dust and gas) began to fragment and contract, probably due to the shock waves caused by one or more supernova explosions.

Supernovas are massive stars that have reached the end of their lives. They explode either completely or partially and expel an enormous quantity of energy and a variety of chemical elements into space. They also produce shock waves that create ripples in the structure of space like the ones that break the surface of a pond when a stone is thrown into it. When these irregularities come into contact with gas or dust in space, they produce significant perturbations in the interstellar material.

This fragmented the original nebula into smaller clouds, each of which began to rotate as a result of the shock sustained. Under the force of its own gravity, one of these clouds—the one which would eventually form the solar system and which we will call the protosolar nebula— began to contract more and more. By the law of conservation of angular momentum (the same one that makes a skater spin faster when she wraps her arms around her body), the cloud's rate of rotation sped up as it got smaller. This in turn produced a flattening of the cloud by cen-

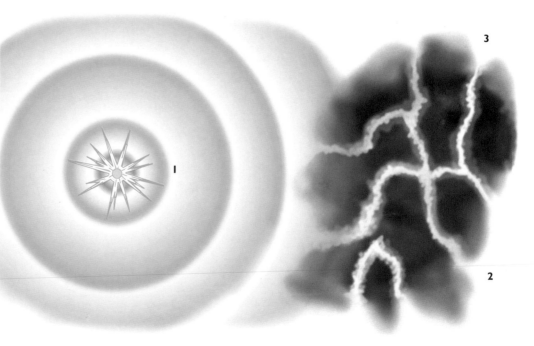

Above, the center of the Orion Nebula, photographed by the Hubble Space Telescope, is the seat of continual star-forming activity such as that which characterized the birth of our solar system. (C.R. O'Dell, Rice University)

Above right, shock waves from the explosion of a nearby supernova (1) may have catalyzed fragmentation (2) of the cloud (3) from which the protosolar nebula originated.

Facing page, the explosion of Supernova 1987A in the Large Magellanic Cloud. (David Malin/AATB)

trifugal force. Still under the force of gravity, most of the material was deposited at the center of the cloud, where, at a certain point, having reached sufficient pressure and therefore a high enough temperature (about six million degrees), the nuclear reactions that led to the creation of our Sun became possible.

The rest of the cloud, meanwhile, quickly became a relatively thin disk. Dust particles along the plane of the disk continued to collide with each other at an ever-increasing rate. Through this action, as well as the concomitant magnetic and electrostatic forces, the particles began to

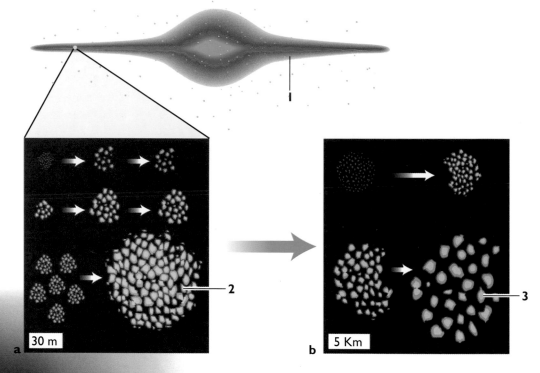

30 m

5 Km

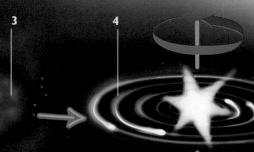

*Stages in the formation of the solar system:*
1. *Original protosolar nebula*
2. *Gravitational contraction*
3. *Birth of the Sun*
4. *Formation of the protoplanetary disk*

*Top, first steps in the formation of terrestrial planets by means of collisional aggregation:*
1. *Protosolar nebula*
2. *Granular accretion*
3. *Planetesimals*

coalesce, forming granular masses about a meter in diameter. At this point, gravitational forces gained the upper hand, and the coalescing material began to pull together ever more forcefully. This caused even more sizable objects to form, with dimensions on the order of a few kilometers. Called planetesimals, these objects orbited the Sun on trajectories with various inclinations and eccentricities. At the same time, the young Sun began to blow a strong "wind" loaded with charged particles (protons and electrons), which almost completely stripped the inner part of the solar system of gas.

Then a kind of planetary billiard game came into play. Planetesimals on intersecting trajectories sustained collisions that were destructive if they happened at high speed and aggregative if they took place at a lower velocity. The latter prevailed, and in the end, bodies about the size of the Moon were formed (roughly 3,000 kilometers in diameter) that orbited in barely inclined and not very eccentric trajectories. These survivors of the game eventually grew to become planets of the "terrestrial" type—Mercury, Venus, Earth and Mars—and wound up in stable, nearly circular and only slightly inclined

orbits that kept them from colliding with each other. This process probably lasted for 100 million years for the Earth, a little less for Venus and Mercury and a little longer for Mars.

Just beyond Mars, the planetesimals could not coalesce to form a planet because they were affected by Jupiter's enormous gravitational forces. The presence of this giant planet, 318 times bigger than the Earth, had two distinct effects. First, it hindered the accretion of dust into larger masses, hurling it off in every direction—toward the Sun or out of the solar system. Finally, it prevented any additional aggregation of the planetesimals that had been able to form.

The result was that between Mars and Jupiter, a large number of big chunks of rocky material, called minor planets or asteroids, remained—perhaps as many as 100,000 (just over 6,000 have been identified so far). The word asteroid means "starlike," because through a telescope, these objects appear indistinguishable from stars. Their diameters range from just a few meters up to 1,000 kilometers for the largest one, called Ceres (which was also the first one discovered, in 1801).

It is thought that farther from the Sun, beyond Mars, there was a different mechanism for the formation of planets. Here, the influence of the solar wind diminished considerably, and the gas of the protosolar nebula essentially remained *in situ*. Gravitational perturbations by the Sun caused some rings of gas to detach from the outer regions of the nebula and slowly condense into large, roughly spherical masses called gaseous protoplanets. These then led to the formation of Jupiter and Saturn, enormous planets with about the same chemical composition as the Sun, mainly hydrogen and helium. The model of collisional aggrega-

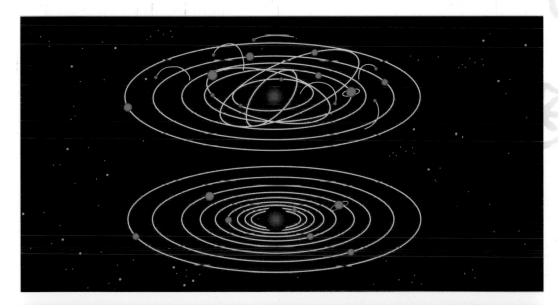

*Top, second stage of terrestrial planet formation.*

*Bottom, formation of large gas planets.*

tion of planetesimals, however, does seem to explain the origin of the more distant planets Uranus and Neptune, except that these were formed not only of rocks but also of various ices—water, carbon dioxide, ammonia and methane—that condensed due to their great distance from the Sun and the resulting low temperature. What we earlier called planetesimals can now assume their proper name—comets. The protosolar nebula was less dense in this region, and the embryos of Uranus and Neptune moved around the Sun far more slowly than did Jupiter and Saturn. For this reason, they were not able to collect as much primordial hydrogen and helium as the two giant planets, and their size bears witness to their intermediate status between the giants and the terrestrial planets.

## The Edgeworth-Kuiper Belt

Only 20 years ago, when its moon Charon was discovered, did we learn that Pluto is so small—2,320 kilometers in diameter, with only one-sixth the mass of our Moon—and that it seems more like a moon itself than a planet. Indeed, for a long time, we thought Pluto might be an escaped satellite of Neptune that had established an independent orbit around the Sun following a collision with some vagrant object. Recently, however, the hypothesis has been advanced that like Uranus and Neptune, Pluto and Charon (half the size of Pluto) formed from the collisional aggregation of frozen planetesimal fragments. This left as traces not only these two objects but thousands of frozen planetoids located beyond Pluto in the so-called Kuiper belt, also known as the Edgeworth-Kuiper belt after the two astronomers who proposed its existence between 1949 and 1951.

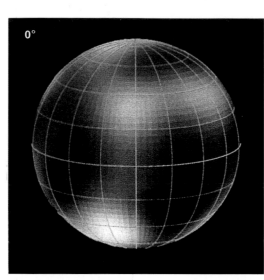
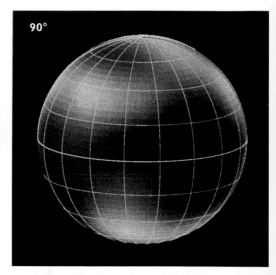
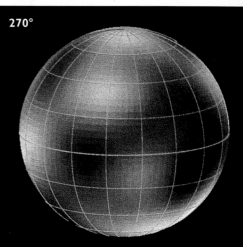
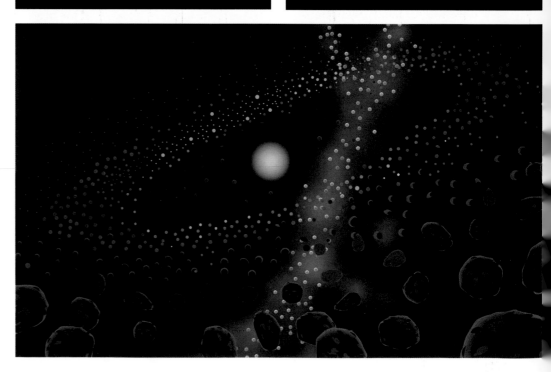

Top, four hemispheres of Pluto spaced 90 degrees apart in reconstructions made from images taken by the Hubble Space Telescope. (A. Stern, M. Buie, L. Trafton/ESA-NASA)

Bottom, artist's conception of the Edgeworth-Kuiper belt, a reservoir of short-period comets and "plutinos."

zone, which leads us to suppose that the total population of the Edgeworth-Kuiper belt may amount to several billion. All the short-period comets seem to originate there.

In June 1997, Luu, Jewitt and others an-

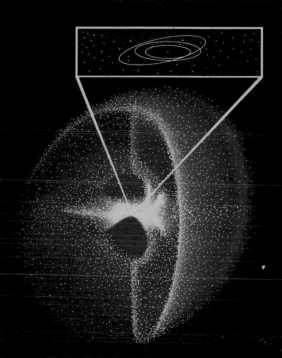

According to Alan Stern of the University of Colorado, the remaining planetoids, with diameters around 1,000 kilometers, never got the chance to form true planets because they were pushed to the outer edges of the belt, leaving behind only Pluto, Charon and Neptune's largest moon Triton (2,700 kilometers in diameter). In any case, small planetoids under 500 kilometers in diameter would have had to stay in the belt. This theory has recently received resounding observational confirmation. Since 1992, about 40 objects 100 to 400 kilometers in diameter have been discovered beyond Neptune. They are believed to be some of the biggest and closest members of the Kuiper belt. According to Jane Luu of Harvard University and David Jewitt of the University of Hawaii, there should be at least tens of thousands of bodies over 100 kilometers in size. They have been nicknamed "plutinos," meaning "little Plutos." A few years ago, by pushing the Hubble Space Telescope to the extreme limit of its instrumental capabilities, numerous smaller objects with diameters from a few to about 10 kilometers were discovered in this

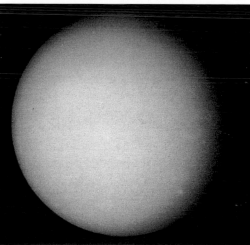

*Uranus (above) and Neptune (top left) photographed by the Voyager 2 probe in 1986 and 1989, respectively.*

*Top right, artist's conception of the Oort cloud, the source of all long-period comets. Inset diagram shows orbits of the outer planets that are otherwise invisible within the scale of the main drawing.*

35

nounced in the journal *Nature* the discovery of the category's biggest member, designated 1996TL$_{66}$, a 500-kilometer-diameter body with a highly eccentric orbit that carries it 132 AU from the Sun. The discovery would lead us to believe that other bodies like this one reside in the immense space between the Kuiper belt and the Oort cloud (see next section). It also leads its discoverers to hypothesize that the Kuiper belt is far larger and contains far more mass than was originally supposed.

## The Oort Cloud

Comets, then, are like the frozen planetesimals that, as we have already seen, led to the formation of the planets Uranus and Neptune and probably Pluto and its moon Charon as well. A surprising confirmation of this scenario has come from analysis of data from the Voyager 2 probe. According to W.B. Hubbard of the University of Arizona, Uranus and Neptune are composed of an undifferentiated mass of billions upon billions of comets, rather than having three internal layers like those of Jupiter

*Above, artist's conception of March 1986 encounter between Halley's Comet and ESA's Giotto probe.*

and Saturn—a solid nucleus, a liquid mantle and a thick atmosphere.

After the formation of Uranus and Neptune, the remaining planetesimals (or comets) had fairly circular orbits and distributed themselves within the Edgeworth-Kuiper belt. Later, however, influenced at length by gravitational perturbations from the embryonic Uranus and Neptune, the size and eccentricity of their orbits increased. At this point, the gravitational influence of the dust, gas and stars distributed along the plane of our galaxy made itself felt. These perturbations caused the comets to move even farther out, to an average of 1,500 billion kilometers from the Sun, into what is known as the Oort cloud. An enormous shell composed of thousands of billions of comets,

the Oort cloud was first suggested in 1950 by Dutch astronomer Jan Oort.

From there, continual galactic perturbations pushed many comets as far as 7 trillion to 20 trillion kilometers from the Sun, into the farthest reaches of the Oort cloud. According to recent estimates, the Oort cloud is an enormous spheroid, centered on the Sun, with dimensions of 30 by 24 trillion kilometers, or 3.2 by 2.5 light-years. The outer region of the Oort cloud is believed to contain 100 to 1,000 billion comets, with 500 to 1,000 billion in the inner region.

Once situated in the Oort cloud, comets can no longer feel the gravitational pull of the planets. But since they are now, in effect, halfway between our Sun and the closest stars, they are subject to the stars' gravitational pull. In fact,

about once every 100,000 years, during the course of its motion around the center of our galaxy, the Sun moves a few dozen light-years closer to some of these stars. This increased proximity, along with ongoing galactic perturbations and occasional encounters with giant molecular clouds, can strip comets away from the Sun's gravity or, alternatively, cause them to fall headlong toward our star.

## A Comet's Coma and Nucleus

When this happens, the affected comet's orbit, which was initially almost circular, transforms to an extremely long ellipse, and the object begins a journey toward the Sun that will last millions of years. If it swings close enough, the Sun's heat will cause the comet's ices to melt, and a body that was previously insignificant on the astronomical scale will then sprout a tail of stupendous proportions.

Regarding the size of a comet's nucleus, our only direct information comes from Halley's Comet, which was measured up close by the

*Left, exceptional high-resolution glimpse of Halley's nucleus produced from a composite of at least 68 images taken by the Halley Multicolor Camera on board Giotto.*

*Above, Comet Hyakutake's coma photographed by the author on March 23, 1996, when its diameter was over 250,000 kilometers.*

The quantity of gas and dust expelled from the nucleus varies from one comet to another. For example, a comet with a very short period like Comet Encke (3.3 years) loses 200 kilograms of dust per second; one with a medium period like Halley (76 years) loses 5 tons; and a nonperiodic comet like Arend-Roland loses up to 75 tons. The dust particles are typically between 0.05 and 10 microns in size. Data about Halley's Comet from the Giotto probe suggest that there are two components—lighter elements, such as carbon, hydrogen, oxygen and nitrogen, and heavier ones, such as silicon, magnesium and iron. And while a medium-sized comet like Halley emits about 20 tons of gas per second, a big one like Hale-Bopp expels more than eight times this amount. Again looking at the data collected on Halley, most of the gas is water vapor (80 percent); the second principal constituent is carbon monoxide (10 percent); and the remainder is made up of carbon dioxide, methane and ammonia.

European probe Giotto in 1986, when it had a close encounter of about 600 kilometers from the comet's nucleus. This revealed that the comet's core looks like a big potato, 16 by 8 by 8 kilometers in size. Other comets have been measured indirectly, however, and with few exceptions, which we will discuss later, all of them are between 2 and 20 kilometers in size.

The gases produced when the ices sublimate in the heat of the Sun are ionized by the Sun's ultraviolet radiation, and dragging along any dust that is liberated from the rocky matrix of the nucleus, these produce the so-called coma, which is tens or hundreds of thousands of kilometers in diameter.

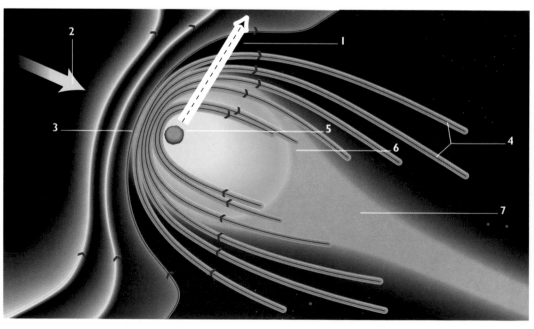

Above, formation of a comet's double tail, beginning with the action of the solar wind and radiation pressure:
1. Orbit of the comet
2. Solar wind and radiation pressure
3. Chaotic magnetic fields
4. Ion tail
5. Dust tail

Above, detailed sketch of interaction between the solar wind and the coma:
1. Orbital path
2. Direction of solar wind
3. Bow-shock waves
4. Magnetic field lines
5. Nucleus
6. Coma
7. Tail

Facing page, the curious shape of Comet Humason's tail, photographed with the big Schmidt camera at Mount Palomar in 1962. This comet had the stature of a Great Comet, but its perihelion was quite far from the Sun, at 2.13 AU, so it barely reached naked-eye visibility.

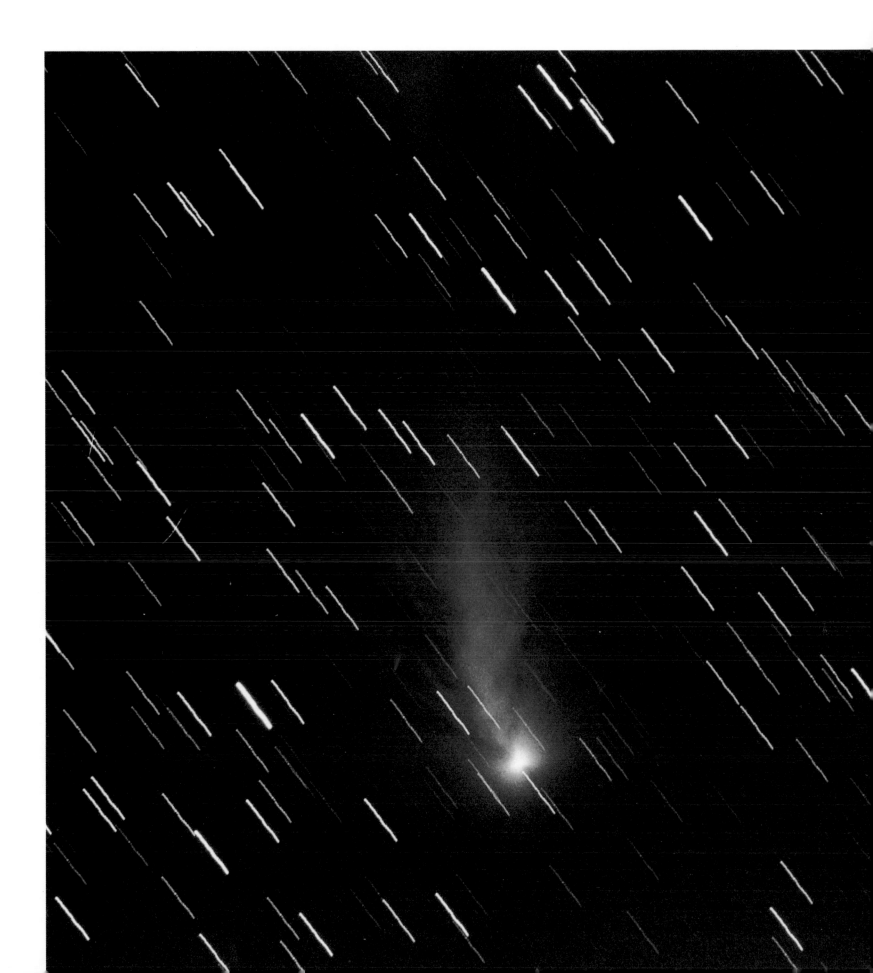

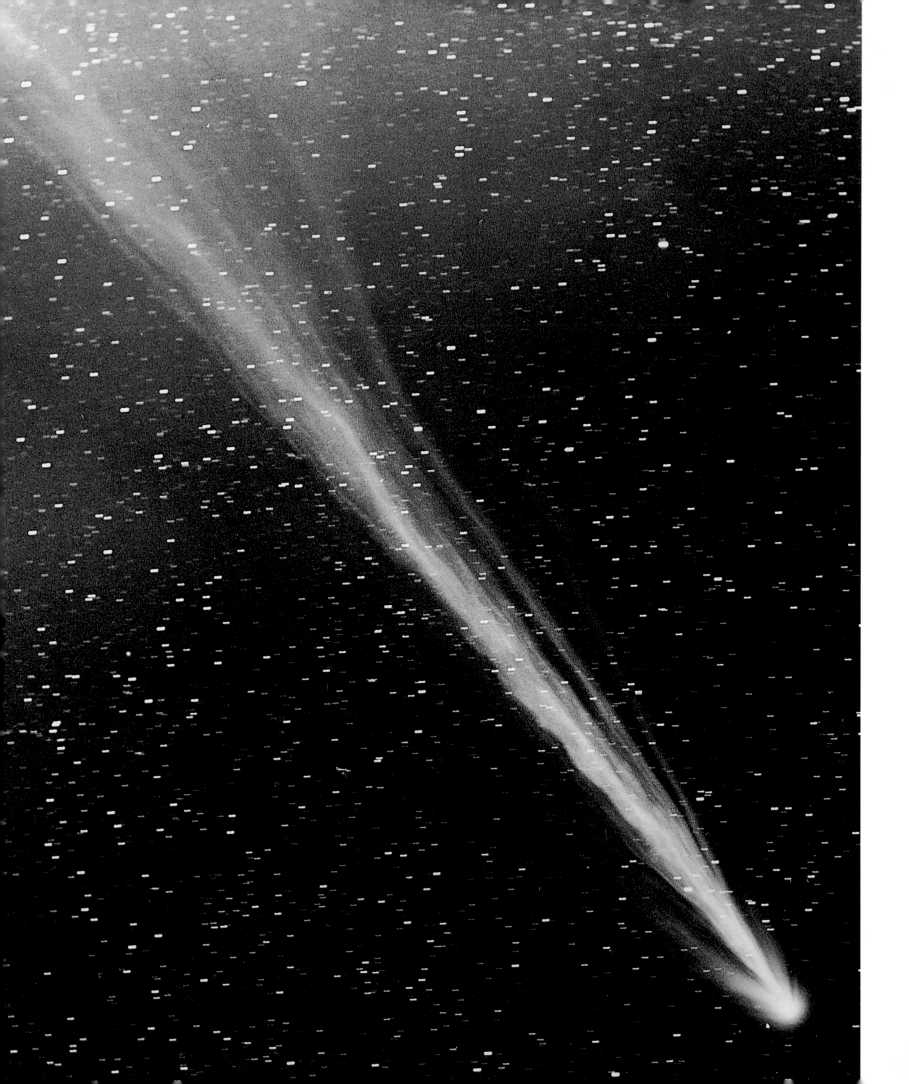

## Two Tails

If the comet gets even closer to the Sun, it forms a tail, or even two separate tails—one of dust and the other of ionized gas (plasma). The dust is propelled away by the pressure of sunlight and, as we saw in the previous chapter, forms a curved tail, because its lighter particles are pushed farther than the heavier ones. Continuing to move along the comet's orbital path, the lighter particles then assume wider orbits than the heavier particles, and according to Kepler's laws, they thus move more slowly and fall behind with respect to the heavier particles. The yellowish color of the dust tail is simply due to the dust reflecting sunlight.

The ionized gases are instead propelled by the solar wind, a stream of protons and electrons emanating from our star. The gases in the coma collide with the solar wind, producing a curved shock wave that travels 50,000 to 100,000 kilometers ahead of the nucleus as a kind of bow wave, like the one generated by a ship moving through water. Ionized particles are channeled along this wave, following the lines of force of the Sun's magnetic field. The solar wind, traveling through interplanetary space at 400 kilometers per second and having a very low density of no more than 10 particles per cubic centimeter, does not disturb the neutral atoms and molecules of the coma, but it does impel the charged plasma particles it encounters to interact with and follow the lines of force of the Sun's magnetic field. The comet's plasma tail shines because its atoms absorb and then reemit sunlight at various wavelengths by a process of fluorescence. The human eye is most sensitive to the light emitted by ionized carbon monoxide. This is at the wavelength of

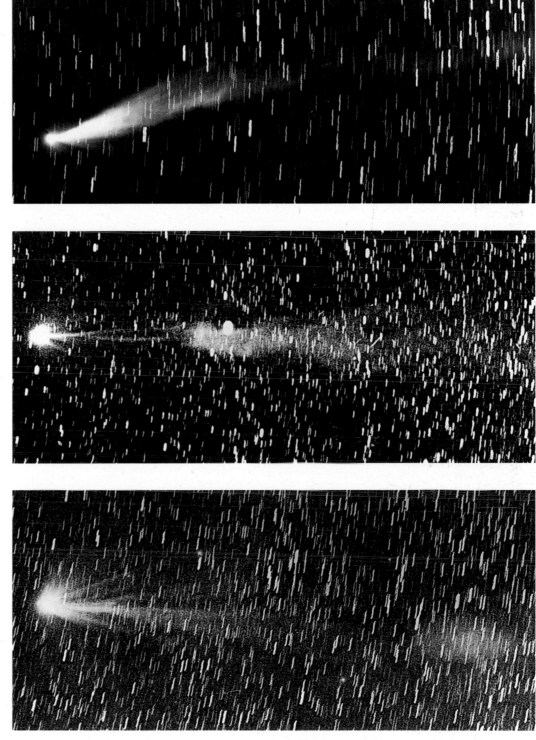

Facing page, Comet Morehouse, photographed by Barnard at the Yerkes Observatory on November 15, 1908. Even today, after so much time has passed, this comet is still one of the best examples of the peculiar structures that can form in the plasma tail.

Above, Comet Morehouse was also the first documented case of detachment of the ion tail.

In these sequential images taken on the nights of September 30, October 1 and October 2, 1908, the tail can be seen separating and moving away from the coma.

the color blue, which is why the plasma tail appears this color.

Even though it is more evanescent than the dust tail, the plasma tail achieves a greater length, often over 100 million kilometers. Since a comet can move in such a way as to cut across the path of the solar wind, its plasma tail is often curved, somewhat like smoke leaving a chimney and coming into contact with air currents. Indeed, because the Sun's magnetic field fluctuates greatly and continuously, the

plasma tail can generate highly contorted and complex shapes, including rays, clusters, streaks, nodules and spiral structures. These move very rapidly along the tail at speeds reaching 50 kilometers per second; they can produce visible variations in the tail in as little as half an hour. The first to document these phenomena clearly was Edward Emerson Barnard, one of the pioneers of astronomical photography. In the marvelous photographs he took during the late 1800s and early 1900s, using

simple instruments of limited focal length, the extraordinary complexity of plasma tails comes into full view.

Barnard's photographs, taken at the Yerkes Observatory, also documented for the first time the phenomenon of the detachment of the plasma tail—when it breaks away from the coma and floats off into space, to be replaced by a new tail. The most striking displays of this phenomenon during those years were exhibited by Comet Brooks in 1893, Halley in 1910,

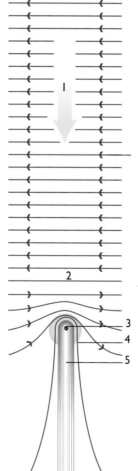
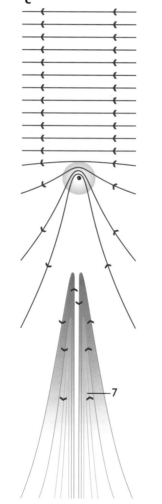
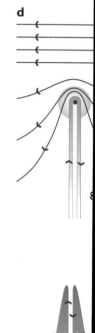

*Left, extraordinary sequence showing the detachment and withdrawal of the plasma tail from Comet Hyakutake on the night of March 24/25, 1996, taken by Eraldo Guidolin. Exposures were made about 20 minutes apart, following Barnard's suggestion in a 1905 essay that a shot be taken every half-hour to document the hourly history of mutations in the plasma tail and to determine the velocity of the gas particles. By analyzing these images, for example, we may deduce a velocity of about 40 kilometers per second.*

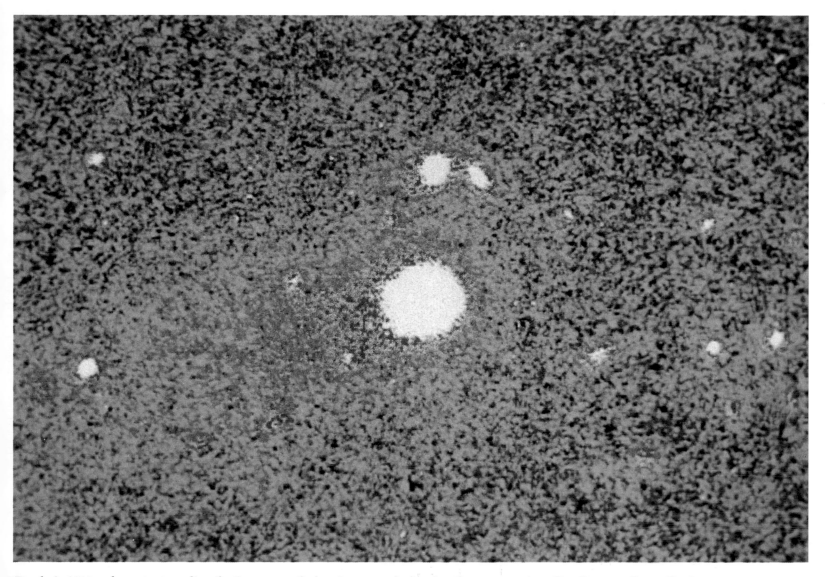

Brooks in 1911 and, most extraordinarily, Comet Morehouse in 1908.

More recently, other famous comets have exhibited the same phenomenon. Among the best were Mrkos (1957), Kohoutek (1974), Halley again (1986) and Hyakutake (1996). We now know that this phenomenon occurs when a comet crosses areas of the Sun's magnetic field that are of opposite polarities. A comet loses its tail when it crosses the border of one magnetic sector into a new sector having opposite polarity to the one in which the tail was formed. The Sun's magnetic field consists of four sectors of alternating polarity expanding through space as they rotate with the Sun over a period of 25 days (3.6 weeks). Consequently, comet-tail separation can occur an average of slightly more than once a week.

## Do Comets Burn Out?

When we consider that comets are quite minuscule objects on the astronomical scale yet are able to produce enormous appendages, we begin to wonder whether comets quickly use up their material. In fact, the material that forms the coma and the tails is destined to be lost in space sooner or later. This is especially true of short-period comets; already smaller,

*Facing page, diagram, sequence of events leading to detachment of the plasma tail:*
*1. Solar wind*
*2. Magnetic sector boundary*
*3. Nucleus*
*4. Coma*
*5. Visible plasma tail*
*6. Magnetic sector boundary*
*7. Detached tail*
*8. New tail*

*Above, the asteroid Chiron, discovered in 1977, was the first representative of a new category of objects that populate the outer solar system, probably including megacomets. (Jane Luu and David Hewitt)*

they lose far more mass by coming close to the Sun more frequently. A short-period comet's coma and tails average a combined mass of only 300,000 tons, while the nucleus of an average comet like Halley has a mass of 100 billion tons. It has been calculated that on each pass by the Sun, Halley loses something like 100 million tons of matter altogether, which amounts to no more than a few meters melted off its surface. So the life span of such a comet can be fairly protracted—no fewer than a thousand passes—though ultimately brief compared with that of the planets and the Sun. However, it is possible for a comet to become inactive in a shorter length of time if its surface is covered by a mass of dark dust that shields its ices and thus reduces the effects of solar forces. As the Giotto probe revealed, the matter spewing from the surface of Halley's Comet comes from only two or three jets. The rest of the surface, a good 90 percent, is inactive and covered by a very dark crust made up of materials that completely resist the Sun's heat and are probably composed of complex organic compounds.

## Megacomets

Many of the comets that come into view are not from the outer edges of the Oort cloud but are from its inner regions or from within the Edgeworth-Kuiper belt, where they are still partially subject to the gravitational pull of the outer planets. Perturbations by the planets modify the comets' orbits, very gradually pulling them into the inner regions of the solar system. These processes lead to the birth of periodic comets, whose orbital period is under 200 years.

As a rule, these comets do not have "the right stuff" to become Great Comets. It seems, however, that a particular group of periodic comets, the so-called Centaurs, may be undergoing a very interesting evolution. The primary representative of this category is Chiron, 180 kilometers in diameter, discovered in 1977 orbiting between Saturn and Uranus. At first, Chiron was considered to be an asteroid, but as it got closer to the Sun, a tenuous coma and a two-million-kilometer tail were observed.

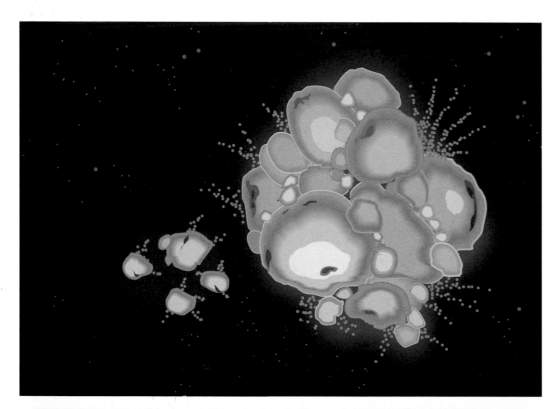

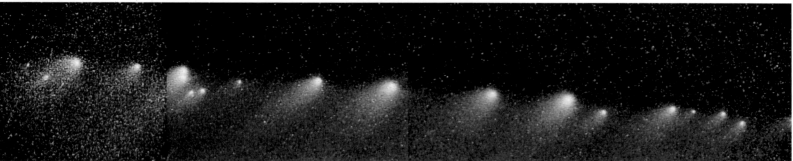

*Top, the "primordial rubble pile" is the model currently in favor for comet nuclei.*

*Above, 20 fragments of Comet Shoemaker-Levy 9 photographed by the Hubble Space Telescope before they collided with Jupiter.*

Pholus, another object of about the same size, was discovered in 1992. Its orbit is even longer than Chiron's. Four smaller Centaurs, with diameters from 30 to 70 kilometers, were discovered between 1993 and 1995. They represent a population of at least a few hundred big comets that have left the Kuiper belt in response to gravitational perturbations, mainly from Neptune. Their orbits currently intersect those of the giant planets, but they are highly unstable. With the passage of a few million years, they could assume even tighter orbits and thus become highly active comets. Future generations may then witness the appearance of comets so bright and spectacular as to exceed the limits of the imagination.

## Modeling the Nucleus

In 1950, American astronomer Fred Whipple proposed a famous theory describing comets as something like "dirty snowballs," very loose aggregations of water, ammonia, methane and carbon-dioxide ices mixed with dust that originated when the solar system was formed. This model has received substantial corroboration and appears quite accurate in light of Giotto's close-up analyses of Comet Halley.

More recently, in 1986, Paul Weissman of the Jet Propulsion Laboratory proposed a better model for the cometary nucleus, calling it a "primordial rubble pile." In this model, the nucleus would be not a single body but a mass of smaller fragments held together weakly by nothing more than their own mutual gravity. The newer model was inspired by—and explains rather well—such cometary phenomena as "outbursts" (paroxysmal flares in activity of the nucleus) and fragmentation of the nucleus, like that of the famous Comet Shoemaker-Levy 9 before it crashed into Jupiter in July 1994.

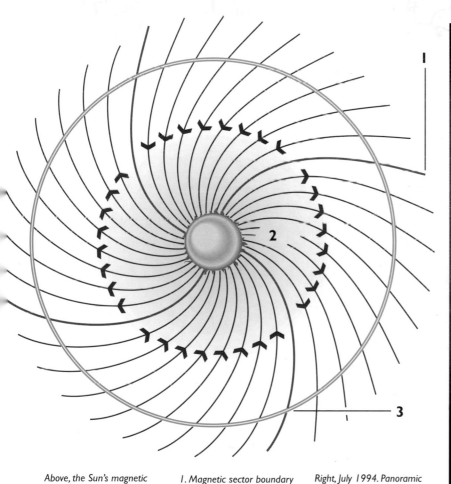

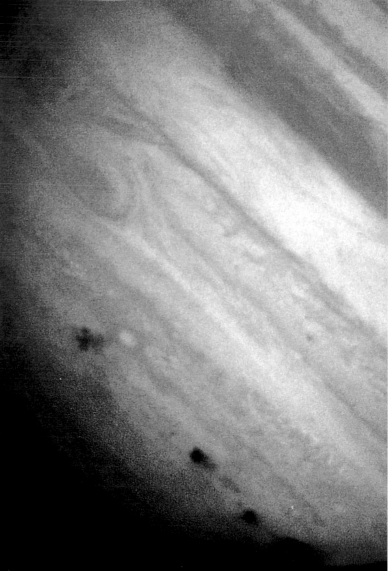

*Above, the Sun's magnetic field (seen here from above the planets' orbital plane) is divided into four sectors. Its rotation with the Sun gives the field a spiral form. Adjacent sectors of the field have opposite polarities.*

*1. Magnetic sector boundary*
*2. Sun*
*3. Earth's orbit*

*Right, July 1994. Panoramic view of scars left on Jupiter after collision with the biggest fragments of Comet Shoemaker-Levy 9. (HST Comet Team/NASA)*

45

# Observing the Great Comets

## Discovering Comets

Discovering comets has a fascination all its own, because it is the only kind of astronomical discovery where the celestial body detected is christened with the name of its discoverer (or discoverers). In the past, up to three independent discoverers were acknowledged, but now only two per comet are recognized.

Discoveries in the 19th century were made exclusively by visual means, employing both reflecting telescopes (like Brooks, who used telescopes 13 and 23 centimeters in diameter) and refractors (like Barnard and Swift, who used

15-centimeter instruments), usually with a long focal length. In contemporary professional circles, however, wide-field photographic telescopes are used or semiautomatic electronic CCD cameras, like the one connected to the 91-centimeter Spacewatch telescope at Kitt Peak Observatory in Arizona. Some comets are even discovered by artificial satellites. The Solwind satellite, dedicated to the study of the solar wind, discovered six comets of the type that graze the Sun, while the Solar Maximum Mission (SMM) satellite discovered 10 very small ones before they entered the Sun's atmosphere. One satellite discovery was shared with human observers. In 1983, the infrared satellite IRAS shared the honor with Araki of Japan and Alcock of Great Britain. Thanks to its extremely close approach to Earth, at less than five million kilometers, this comet was, among other things, the most brilliant of the 1976–96 period. Among

comets recently discovered by satellites are those found by the ESA-NASA probe SOHO, which photographed a few dozen Sun-grazing objects while it was orbiting the Sun.

A lot of comets are still discovered by visual observation by a throng of amateur astronomers all around the world. Many still prefer to observe with classic refractors, but with a short focal length, like the late Leslie Peltier, an American who was active around the middle of the 20th century and used a 15-centimeter f/8 refractor, or Australian William Bradfield, who is the 20th century's world leader for visual discoveries. Bradfield, who uses a 15-centimeter f/5.5 refractor and was a professional engineer in the field of jet propulsion (now retired), possesses another unique record: he is the sole discoverer of all of his 17 comets, which therefore carry his name alone. Some, like American David Levy, have a predilection for big reflec-

*Above, one of the brightest comets discovered by Australian amateur astronomer William Bradfield appeared in autumn 1987, here photographed by the author.*

*Right, Comet IRAS-Araki-Alcock, codiscovered by a satellite (IRAS) and two amateur astronomers, reached second magnitude in the spring of 1983. (Courtesy of C. Comovici and S. Ortolani, Asiago Observatory)*

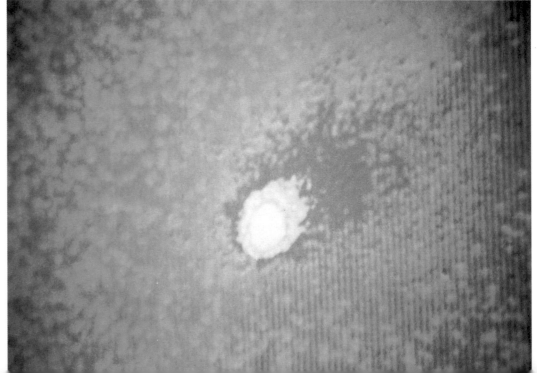

### THE TOP-RANKING COMET DISCOVERERS

| Discoverers | Discoveries | Country | Active Period |
|---|---|---|---|
| Carolyn Shoemaker | 32 | USA | still active |
| Eugene Shoemaker | 31 | USA | 1983–1997 |
| Jean Louis Pons | 30* | France | 1801–1827 |
| David Levy | 21 | USA | still active |
| William Brooks | 21 | England | 1885–1911 |
| William Bradfield | 17 | Australia | still active |
| E.E. Barnard | 16 | USA | 1881–1899 |
| Lewis Swift | 14 | USA | 1878–1899 |
| Charles Messier | 13 | France | 1759–1798 |
| Antonin Mrkos | 13 | Czech Republic | 1948–1996 |
| Wilhelm Tempel | 13 | Germany | 1859–1877 |
| Minoru Honda | 12 | Japan | 1947–1968 |
| Michel Giacobini | 12 | France | 1878–1907 |
| Malcolm Hartley | 11 | USA | still active |
| Alphonse Borrelly | 11 | France | 1873–1912 |
| Leslie Peltier | 10 | USA | 1930–1954 |
| Friedrich Winnecke | 10 | Germany | 1854–1877 |

*four unconfirmed; only 26 carry his name

tors; he uses a 40-centimeter telescope for his visual discoveries (he also collaborated on photographic research with the Shoemakers at Mount Palomar).

Many amateurs, however, find it more amenable to use giant binoculars that permit them to exploit a wider field of view and to experience more comfort while carrying out a very stressful research operation. Among the first to use this type of instrument were astronomers at the Skalnate Pleso Observatory, founded in 1943 by Antonin Becvar in the Tatry Mountains of Czechoslovakia, at 1,780 meters. Between 1946 and 1959, at least 18 comets were discovered by five different observers using 25x100 binoculars. One woman stands out among them, Ludmilla Pajdusakova; she was a solar astronomer by training and found six of the 18. Antonin Mrkos figures among this group as well. He continued independently and has collected at least 13 trophies. His most

important discovery has a curious history. By naked eye, he saw the comet's tail rise from the horizon before dawn on August 2, 1957. The coma sprouted from the horizon with the sky already bright. In fact, the comet had been discovered in Japan on July 29 and in the United States two days later, but Mrkos was the only one who followed the correct protocol in these cases by immediately sending a telegram to the International Astronomical Union. Thus the comet carries only his name.

George Alcock of Great Britain, who has discovered six comets, uses 25x105 binoculars. The Japanese, often using gigantic 25x105 binoculars, are currently among the most prolific discoverers (just remember the names of Masaru Arai, Tsuruhiko Kiuchi, Tsutomu Seki, Minoru Honda and Kaoru Ikeya).

As the above table shows, Carolyn Shoemaker of the United States is still active and considered to be the greatest discoverer of

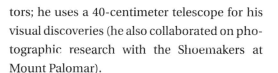

*In the above table, the all-time greatest comet discoverers are listed along with the number of their discoveries, country and active period.*

*Top, the husband-and-wife team of Eugene (right) and Carolyn Shoemaker and teammate David Levy (left) immortalized beside the 46-centimeter Schmidt telescope at Mount Palomar that they used to make most of their discoveries. (Alan Levenson for Time)*

*Above, 1957 was an auspicious year. In addition to Comet Arend-Roland (mentioned in Chapter 1), Comet Mrkos also appeared and almost reached the proportions of a Great Comet, achieving a magnitude of 1 and a tail 13 degrees long. (Observatoire de Haute-Provence du CNRS)*

comets ever, in part by using a highly fruitful photographic method with the 46-centimeter Schmidt telescope at Mount Palomar. Among other women who are famous as historical discoverers is Caroline Herschel, sister of the great William Herschel. Using William's big reflecting telescopes, she was the first woman to discover a comet; between 1786 and 1798, she discovered eight comets. American astronomer Maria Mitchell detected seven comets. Currently, the United States' Jean Mueller holds a preeminent position: She has discovered eight comets as well as nine remarkable asteroids and at least 49 supernovas!

Italians, on the other hand, have never paid much attention to this pastime. You need only the fingers of both hands to count the comet discoverers in Italy. Six of the discoveries are attributed either to the already-cited Giovan Battista Donati, director of the astronomical observatory in Florence and active between 1854 and 1864, or to the Jesuit priest Francesco De Vico, director of the Collegio Romano Observatory, who made his discoveries in just three years, from 1844 to 1846.

Lorenzo Respighi, director of the Bologna Observatory, discovered three comets between 1862 and 1863. Father Angelo Secchi, De Vico's successor as director of the Collegio Romano Observatory, found two of them in 1852 and 1853. Temistocle Azona, director of the Palermo Observatory, made a discovery in 1890.

In the 20th century, only three Italians succeeded in this endeavor. Amateur astronomer Giovanni Bernasconi of Lombardy found three between 1941 and 1948, two with 15x60 binoculars and one with a 120-millimeter refractor. Professional astronomer Roberto Barbon discovered a comet at Palomar Observatory in 1966. The most recent was Mauro Vittorio

Zanotta of Milan, whose success came in 1991.

Naturally, the pleasure of discovery is intensified if the found object is destined to become bright or outright spectacular. On the whole, it is not exactly true, as has been written, that it is enough to discover a comet to be "immortalized in the heavens." For your name to be remembered at length or even forever, the newly found comet must become a Great Comet. This is what happened, for example, to Yuji Hyakutake of Japan in 1996.

## An Unexpected Arrival

The discovery of this comet was completely unexpected. Everyone was getting ready to observe Comet Hale-Bopp, which had already been seen in July 1995—almost two years in advance of its passage to perihelion—and now appeared as the first Great Comet in more than 20 years. Instead, the comet discovered by Hyakutake at the end of January 1996 stole the spotlight. It quickly became evident that the new arrival would pass very close to Earth at the end of March. During the nights from March 24 to 27, it would be just 0.1 AU (15 million kilometers) away. Surely, it would shine like a first-magnitude star, although its luminosity would be spread across a rather large area, because it was so close to Earth. While predicting its luminosity was pretty easy (since the comet was already fairly close to the Sun when discovered, it couldn't be a big surprise), there were convoluted predictions on the length and visibility of its tail. Burned by earlier episodes, when some comets had failed miserably to live up to predictions, almost everyone was keeping his cards close to his chest.

At first, this caution seemed warranted. On March 14, the tail was less than 1 degree long, and on March 20, it was barely 2 degrees.

## Planet Earth Flyby

Then, however, the unthinkable happened. In three short nights, the tail stretched out incredibly as the comet approached Earth. On the evening of March 23, we were in the Falzarego Pass at 2,105 meters together with astrophiles from Feltre, Cortina and Conegliano, on an island of serenity above a thick quilt of clouds that blanketed the lowlands and reached up to 1,800 meters. The view was splendid. The sky was absolutely clear of clouds and incredibly transparent. Comet Hyakutake had undergone an awesome metamorphosis. Its tail was 26 degrees long and extended from Bootes to Coma Berenices. It was far brighter than predicted—fully visible to the unaided eye, with filaments and structures of a rare complexity and beauty. The tail, composed of ionized gas, was slim at its source, close to the coma, then spread wider and wider. The dust component, which was wide and overlaid the first part of the gas tail, was only 2 or 3 degrees long. At 0.5, the magnitude of the coma had blossomed, falling somewhere between the brightness of Spica and Arcturus. It was 1 degree across. Finally, we could say it was just what we had hoped for.

It was our first Great Comet, and it was the first in two decades. It was incredible to see our first Great Comet in a sky that black, showing stars beyond seventh magnitude. We remained, awestruck and ecstatic, watching and photographing throughout the night. It was almost annoying when we resorted to our instruments to observe that apparition way up above us. It pressed on us, humbling and magnificent for the whole night; it hacked away 2,000 years of history in one fell swoop by summoning up the emotions and apprehensions of the ancients. And it was best seen with the naked eye. The comet seemed to hang right up there, not mil-

*Facing page, top left, photo of Comet Hyakutake taken by the author on the night of March 23/24, 1996, with a simple 55mm lens.*

*Bottom left, Hyakutake on the night of March 23/24 photographed by Carmelo Zannelli of the Palermo ORSA with a 200mm telephoto lens.*

*Top right, Hyakutake photographed by the author on the night of March 24/25 with a 55mm lens.*

*Bottom right, the author's fisheye lens reveals the great size of Hyakutake's ion tail, which extended over one-quarter of the sky (the two opposite horizons can be seen at upper right and lower left).*

lions of kilometers away in space but right up there like a vapor emanating from the ground, just as the ancients thought. When the Milky Way rose around 3 a.m., we saw one more testimony to the sky's clarity. The star clouds of Cygnus, Scutum and Sagittarius were so dense, it looked as if they could rain.

As day broke, Hyakutake was still high in the sky, brandishing every bit of its power. And this was the comet's most awe-inspiring trait. Other Great Comets have normally been visible at dawn or sunset, but Hyakutake was visible all night, for eight consecutive hours of darkness. We knew that a magnanimous fate had smiled upon us. Other comets have been brighter, others have had a longer tail, but only this one, our first Great Comet, could be seen the whole night, and we knew that it would never be this way again.

The following evening, we were still at Falzarego. The ion tail had grown remarkably compared with the night before and was about 42 degrees long. The first 20 degrees was very bright. By telescope and binoculars, the part of the tail closest to the coma was rich with intricate detail. The second 20 degrees of the tail varied in luminosity from average to faint. Through binoculars and even more in photographs, a great discontinuity was clearly visible in the ion tail. The dust tail was as long as the evening before and somewhat broader but more or less the size of the coma. The coma looked highly diffuse and essentially uniform

in brightness, with a small central condensation that resembled a very bright star. Using a telescope, we could also see a thin, bright jet of material coming out of the false nucleus, or central condensation—a dense clump of gas and dust most recently emitted into the coma. The coma was 1.5 degrees across, the equivalent of about 400,000 kilometers at that distance from Earth. At a certain point, Hyakutake's tail began to exhibit a very curious double curve in the form of an S, similar to ancient draw-

*Bottom left, high-resolution shot of the inner region of Hyakutake's coma taken by Salzburg amateur astronomer Gerhard Grau on the night of March 27/28. In addition to the parabolic appearance of its outer shell, we can clearly distinguish the very bright central condensation from which a long, thin jet of matter flows.*

*Bottom right, Hyakutake photographed on March 27 by A. Ghedina of the Cortina Astronomical Association over the dome of the Cortina Observatory, illuminated by the Moon. Below, photo taken with a super fish-eye lens by A. Dimai of the Cortina Astronomical Association on the morning of March 28 at 2,236 meters in Giau Pass in the Dolomite Alps, Italy.*

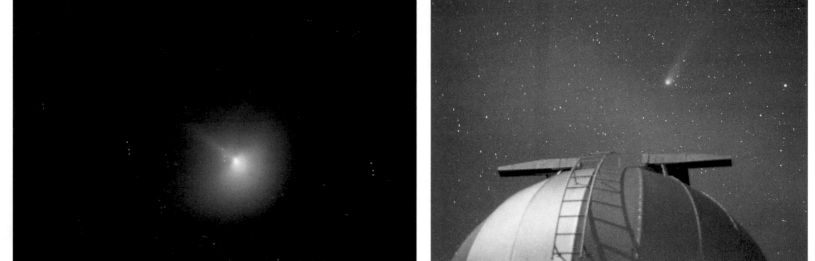

ings in which comets were depicted as lethal signs.

The evenings that followed gave us a stretch of bad weather, although our friends from Cortina persevered incredibly and were rewarded with a few morning clear spots in the high Dolomites. They took more splendid shots and watched the tail grow longer and longer until it reached nearly 50 degrees. The magnitude appeared stable, but it was enough by any measure to place Hyakutake among the century's most spectacular comets. Only six have been brighter, and only one—Halley in 1910—had a much longer tail.

But positively no other comet has been visible under such ideal conditions for the entire night.

## On to the Sun

Once it brushed past Earth, Hyakutake's brightness and long tail diminished, as had been forecast. After April 5, the Moon could no longer interfere with observing, and up until April 20, when the comet became too low to permit further observation, it was still possible to see Hyakutake. It presented a series of highly

*Top, in mid-April, Hyakutake decorated an area of the sky animated by the Pleiades (at bottom), the planet Venus (lower left) and the California Nebula (left of center at top of frame). Photograph by Giuseppe De Donà of the* Feltre Astronomical Association Rheticus.

*Above left, notwithstanding Hyakutake's remarkable distance from Earth—almost 110 million kilometers on April 15—its tail extended 25 degrees, as shown in this image taken with a 50mm lens by Giuseppe De Donà.*

*Above, the contorted appearance of the comet's plasma tail on April 15, photographed by the author with a 300mm lens.*

tripartite, since the ion tail itself branched into two. Naturally, as the comet moved farther and farther from Earth and got closer to the Sun, the real size of its tail grew from one night to the next, because Hyakutake felt the solar wind more forcefully every day. Around April 20, the gas tail reached 50 million kilometers in length. This is an astonishing size, especially if you consider the comet's reduced nucleus. Indeed, NASA's 70-meter antenna at Goldstone was used to advantage during Hyakutake's close visit. By using radar, the diameter of a comet's nucleus was measured with precision from Earth for the first time ever. It turned out that Hyakutake's nucleus was somewhere between one and three kilometers in diameter, about one-fifth the size of Halley's. For it to display so much activity, we must assume that a high percentage—50 percent or more—of Hyakutake's surface is active or able to react to solar forces (by comparison, only about 10 percent of Halley's nucleus is available).

The solar probe SOHO made a second important discovery. It caught up with Hyakutake after it passed perihelion on May 1 and was no longer visible from Earth. A coronagraph on board detected the presence of a third tail made up of heavy dust that barely feels the pressure of solar radiation and therefore tends to diffuse all along the orbit of the comet.

But the most sensational discovery came when the German ROSAT satellite detected x-rays emitted by the comet. X-rays are usually produced by extremely dynamic events that are not easily found in comets. What's more, Hyakutake's emission was very intense. The most likely explanation for the phenomenon seems to be that the gas around the comet's nucleus absorbs solar

intriguing phenomena and spectacular configurations with other celestial bodies.

The coma was now no brighter than magnitude 2.5 on the best nights, the tail no longer than 15 degrees to the naked eye, 22 degrees through binoculars and 25 degrees in long-exposure photographs.

On April 15, due to a sudden release of material from the nucleus, Hyakutake had another little outburst that made it considerably brighter. Observing by telescope, it looked as if the false nucleus was flattened like a pancake, with two winglike structures that got lopped off and

ended up in the dust tail. In photographs taken that night, an irregular contorted structure, which seemed to wrap around it, could be seen in the ion tail. After April 16, the geometry of observation also permitted a view of the unmistakable separation between the plasma and dust tails. Their appearance was quite evocative. They formed a kind of swallowtail, with the dust tail on the left and the ion tail—characterized by the presence of many threadlike branchings—on the right. The sight was especially beautiful through 20x80 binoculars, and in photographs, you could see that the tail was

*In this photo taken by the author on April 16, we see a clear separation between the white dust tail below and the two branches of the blue plasma tail on top.*

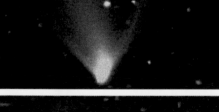

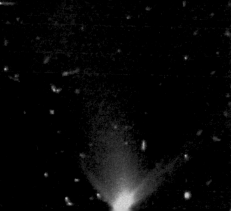

x-rays and reemits them by fluorescence (the same phenomenon that makes the plasma tail shine in the visible spectrum).

## Hale-Bopp Arrives

Unlike Hyakutake, Hale-Bopp's arrival as a Great Comet had been announced long before its passage to perihelion. Americans Alan Hale and Thomas Bopp discovered the comet on July 22, 1995. Even though it was well beyond Jupiter, more than 6 AU from Earth and 7 AU from the Sun, it was, at 10th magnitude, visible even in amateur telescopes. This meant that it was either a very big comet or unusually active or both.

As we have already mentioned, it is not easy to measure the diameter of a comet's nucleus, since what the telescope sees is not the "naked" nucleus but something much bigger—the coma—produced by the release of dust and gas under the action of solar heat and wind. The nucleus remains inactive (and therefore naked) only at a great distance from the Sun on a first approach. The distance is variable from comet to comet according to the quantity and type of volatile material available. So it is also impossible to extrapolate the diameter of a comet by comparing it with the activity of other comets seen at the same distance, because each one reacts in its own way to the activating forces of the Sun.

Indirect estimates suggest that the biggest diameters for various comet nuclei to date would be around 100 to 120 kilometers. Examples include the comet of 1106, already mentioned, and Comet Sarabat, which appeared in 1729. The latter was not especially brilliant, but it was visible to the naked eye, even though it never

got closer than 4 AU (600 million kilometers) from the Sun and 3 AU from Earth. Its absolute magnitude (an index of intrinsic luminosity determined by imagining the body to be placed at a standard distance of 1 AU from both the Earth and the Sun) was estimated at –3, higher than almost any other known comet and 1,500 times brighter than a typical Great Comet.

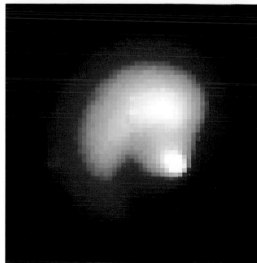

*Left, this sequence of images by the SOHO probe's coronagraph shows the evolution of Hyakutake's triple tail on May 2, 3, 4 and 5. On the right, the plasma tail; the dust tail is in the center; at left, a newly discovered tail composed of heavy dust. (M.D. Andrews and S. Passwaters)*

*Above, Comet Hale-Bopp photographed by the Hubble Space Telescope on October 5, 1995, when it was a billion kilometers from the Sun. Top, the original image; bottom, this enhanced image shows a jet of matter already emerging from the nucleus, surrounding the core with a spiral structure that follows the direction of spin of the nucleus. (H.A. Weaver and P.D. Feldman)*

Hale-Bopp clearly does not measure up to the comet of 1729. At double the distance of the 1729 visitor when it was discovered, Hale-Bopp should have been at least seventh magnitude if it were comparable. Since it was only 10th magnitude, its dimensions are clearly inferior. Its absolute magnitude was very high, however, equal to –1, which is 250 times higher than the average Great Comet. Hubble Space Telescope data yielded an estimate for its diameter of a remarkable 40 kilometers. Moreover, Hubble images taken in October 1995 showed that Hale-Bopp was already very active beyond the orbit of Jupiter. No other known comet has been so active at that distance.

Unfortunately, it soon became clear that the comet would pass rather far from the Sun—0.91 AU—at perihelion, which it reached on April 1, 1997, and that its minimum distance from the Earth, which it reached on March 23, would be especially great, at least 1.31 AU. In other circumstances, Hale-Bopp might have won the title of the most spectacular comet in human memory.

Curiously, Hale-Bopp closely resembles the comet of 1811. It has high intrinsic luminosity, a similar orbital period (estimated at about 21,500 years), a high orbital tilt and a similar distance at perihelion. As we said, the comet of 1811 had a perihelion just beyond the Earth's orbit, and thus, notwithstanding its high intrinsic luminosity, it never reached magnitude zero. At such a great distance, its tail never achieved a greater apparent length than 25 degrees, even though it was truly very long. Nonetheless, its steep orbital tilt kept it well above the Earth's orbital plane, making it visible to millions of people in the northern hemisphere. Though it did not become a monster comet, it was visible to the naked eye for at least 260 days—a record. As we will see, Hale-Bopp was much more poorly positioned for observation, but it unveiled a far brighter self.

## Predictions Confirmed

Naturally, despite Hale-Bopp's promise, many feared a repetition of the fizzled comets Kohoutek and Austin. As it turned out, though, Hale-Bopp was about 200 times brighter intrinsically than Austin and fully 7,000 times brighter than Kohoutek. Among the historic Great Comets, only those of 1106 and 1729 were bigger, and only two others—the comet of 1577 and, again, the comet of 1811—could compare with Hale-Bopp in terms of absolute magnitude. And Hale-Bopp did not disappoint. The spectacle was already visible through ama-

*At least six distinct jets of material appear in this photograph of Comet Hale-Bopp taken in early November 1996 with the author's 25-centimeter telescope.*

teur instruments in July 1996, when the comet began to develop a very wide dust tail that ultimately looked like a long coma. In August, the tail was about 0.33 degree long, and it substantially maintained that length throughout the succeeding weeks. Its real size equaled 8 to 10 million kilometers—not bad, considering the comet was then 3.5 AU from the Sun. Its width was really amazing—about 10 minutes, or almost five million kilometers. Moreover, photographs taken with modest instruments revealed numerous jets emerging from the nucleus, a further indication of activity that had become almost paroxysmal. Even the coma presented an unusual size; it was nearly one million kilometers wide, almost as big as the Sun!

After August, Hale-Bopp began to anticipate its twilight more and more. Eventually, it was visible only during the first part of the night and then, after October, only in the evening. The coma lingered around sixth magnitude from August until October. This is easily explained by remembering that although the Earth approached the comet slightly in the early days of August (2.73 AU), it moved farther along its orbit and reached a distance of 3.05 AU by the end of October.

Another explanation might reside in the fact that when the comet was roughly 3 AU from the Sun (during this time, it was about equidistant from the Earth and the Sun), Hale-Bopp's coma began to exhaust its reservoir of volatile methane, ammonia and carbon dioxide and started to sublimate its water ice.

But things soon changed for the best. At the end of October, Hale-Bopp's brightness began to climb again, reaching fifth magnitude. At the beginning of December, it reached fourth magnitude and was readily visible to the naked eye for the first time. In fact, even though the unaided eye can perceive stars down to sixth magnitude, the issue is somewhat different with comets. Since they are characteristically diffuse objects and thus do not appear as points of light, their brightness is dispersed over a considerable area and consequently seems less. Also, this comet was very low to the horizon, no more than 10 degrees high, so the densest layers of the atmosphere softened its light. The length of its tail, moreover, did not exceed 0.5 degree telescopically (1 degree in photographs)

because of an unfavorable perspective (in fact, its real length was already 15 to 20 million kilometers).

From halfway through December until mid-January, the comet was unobservable because of its apparent proximity to the Sun. It began early-morning apparitions in the middle of January. During that time, the comet brightened to magnitude 3, but the tail seemed to

*Left, panoramic shot taken by the author on the morning of March 2 from Rolle Pass (1,984 meters), in the Dolomite Alps. The peak called Cimon della Pala (3,186 meters) is to the right, lit by the last-quarter Moon.*

*Above, the comet photographed by the author with the 20-centimeter Newton telescope at Feltre Observatory on the morning of February 18.*

55

have stopped getting any longer and appeared visually about 0.5 degree long. The comet was still very low, however, no more than 10 degrees above the horizon.

## The Big Show

Finally, in February, Hale-Bopp began to reveal its true charms. In the early days of the month, under conditions of improved visibility (about 20 degrees high), the comet achieved magnitude 2. Binoculars clearly revealed separate ion and dust tails. While the latter was no more than 0.5 degree long, the gas tail exceeded 1 degree. In photographs, however, their dimensions were at least double those amounts.

In mid-February, the comet finally reached magnitude 1.5. Twenty-two degrees high at dawn, it was now clearly visible to the naked eye, even to a casual observer. Using binoculars under dark skies at an average mountain altitude around 1,000 meters, you could easily distinguish a dust tail more than 1 degree long. This was markedly brighter than the ion tail, which could barely be made out to 4 degrees from the coma. It reached about 7 degrees in photographs, however, equivalent in real terms to about 70 million kilometers.

In March, the comet was visible only in the morning, but its magnitude soared. It was 0.5 on the first of the month, despite a last-quarter Moon. From high altitudes, you could see 2.5 degrees of dust tail and almost 5 degrees of gas tail without visual aids. It grew to magnitude –0.5 by 10 a.m., meeting the most optimistic forecasts and beating Hyakutake by a long shot.

56

*Hale-Bopp photographed on March 7 by Alessandro Dimai and Davide Ghirardo with a 102-millimeter refractor at f/6.*

The double tail finally became very visible to the naked eye. In one week, the dust tail grew to 5 degrees and the gas tail tripled to at least 13 degrees. In mid-March, the dust tail reached 7 degrees, while the ion tail was almost as long. The coma was now magnitude –1.

In theory, Hale-Bopp should have become visible in the evening around March 10. It seems to have been too low, however—under 10 degrees—to be seen, even at locations with a

*Above left, the comet photographed by the author on March 9 with a 300mm lens.*

*Above, the comet photographed by the author on March 16 from Giau Pass*

*(2,236 meters) with a 135mm lens.*

*Left, Hale-Bopp above the Sassolungo peak taken from Pordoi Pass in the Dolomite Alps on the evening of March 13 by Giuseppe De Donà.*

flat horizon. Moreover, the Moon should have been a considerable annoyance after March 12. But thanks to the great brightness of the coma and to skies swept clean by twilight winds, Hale-Bopp trifled with the Moon's interference, shining like a smiling ghost in the northern sky of mid-March. It was an incredible show, like something planned by a shrewd and equally mysterious director. Even from the middle of a city, you could see the dust tail stretching out parallel to the horizon, as if it wanted to make its swirling heavenly path tangible.

At the end of March, the Moon had finally left the sky, and the comet turned into something

from a planetarium show. The dust tail got longer every day, shining brighter and brighter, and even in big cities, it was impossible not to notice the bright coma and the first 4 or 5 degrees of the dust tail. Curiously, the comet's course among the constellations paralleled the Milky Way, so the comet and its two tails appeared to dangle off the star clouds during most of the apparition.

At higher altitudes, the spectacle was outstanding. Even as it grew less and less visible, the plasma tail stretched to 15 degrees (1 billion kilometers)—one of the longest tails ever, in absolute terms. It was embellished with gos-

samer structures that were marvelous to see in binoculars and photographs.

Since early March, the dust tail had displayed some peculiar striations that were slanted greatly with respect to the tail's axis. These structures recalled the synchronous bands observed in several comets in the past, including de Cheseaux, Donati, the comet of 1910, Arend-Roland, Ikeya-Seki and West (see Chapter 1). The striations, however, did not seem to come from dust that was being emitted directly from the nucleus. They may have been the result of secondary processes affecting dust particles in the outer coma. In any event, similar

structures had never been seen in comets so far from the Sun (1 AU). The dust tail got longer and longer and finally measured 12 degrees around April 10.

Near the end of March, the coma already exceeded the apparent size of the full Moon— 0.5 degree, or about two million kilometers across. Moreover, from the first of the month, when observed at high magnification, it displayed a structure of peculiar concentric shells. It shared this aspect with a few comets in the past—Donati in particular. These structures were produced by jets of material emerging from the nucleus, which tend to settle into a spiral form because the nucleus rotates so fast

*Facing page, far left, in this image taken by the author on March 16, the peculiar striations discussed in the text are clearly visible.*

*Facing page, near left, the comet photographed by the author in late March with a 55mm lens; the end of the ion tail fades into the Milky Way.*

*Above, photos taken on March 28 by the author (left) and on March 29 by Giuseppe De Donà (right) demonstrate the progressive fraying of the plasma tail, which ultimately divides into five separate filaments.*

(once every 11.5 hours, more or less). The spiral, viewed somewhat obliquely from Earth, appears to be composed of concentric halos.

Throughout April and for the first days of May, the show went on, generously prolonged by the comet's reservoir of volatile substances. Hale-Bopp stayed at magnitude −1 for nearly a month, from mid-March to mid-April, and remained at a negative magnitude for 50 consecutive days, from March 10 through the end of April. No other Great Comet in history had done as much. It is no surprise that this apparition was the most heavily tracked astronomical event in the history of humanity. And all this at

a time when our planet is shrouded in artificial light that is wasted and poorly directed, illuminating the sky rather than the streets and other public areas. Hale-Bopp outsmarted the city lights and shone forth all over the world. The spectacle rendered a great service in Italy as an ineffable testimony in a long fight against light pollution. On April 5, 1997, in fact, the Italian Astrophiles' Union organized hundreds of star parties called "The Night of the Comet" so that hundreds of thousands of people could observe this extraordinary object from places where artificial lights had been either dimmed or turned off completely.

Hale-Bopp also produced x-ray emissions. In fact, in an effort to go beyond the revelations from Hyakutake, scientists had gone back to the archive of data from the ROSAT satellite (in orbit since 1990) and found another five comets that displayed this peculiarity. This time, however, it was an Italian satellite, called BeppoSAX, that made the discovery, in October 1996.

The most notable discovery relating to Comet Hale-Bopp accrues to yet another Italian, Gabriele Cremonese of the Padua Astronomical Observatory, who, together with his colleagues on the European Hale-Bopp team, was the first to discover a third type of comet

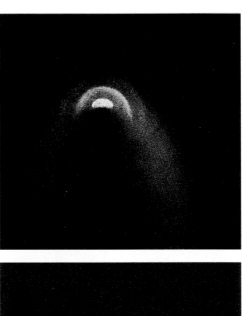

*Near left, bottom, concentric shells observed in Hale-Bopp's coma (photo by the author) are very similar to those observed in Comet Donati's coma, at left, top, in a drawing by G.P. Bond.*

*Far left, Hale-Bopp travels alongside the open cluster M34 on the evening of April 6. Photograph by Alessandro Dimai, Davide Ghirardo and Renzo Volcan.*

*Facing page, top, the comet photographed on April 10 by the author.*

larly difficult these days, since good publications on the subject are easily found at newsstands. Moreover, there are one or more astronomical associations in virtually every city across North America that dedicate time to hosting public observing sessions. You can be sure that all of these associations would organize public observing evenings should another Great Comet make its appearance. This is exactly what happened with Hyakutake and Hale-Bopp.

## Advice for Observers

For those who want to dedicate themselves individually to observing the next Great Comet, there are a few things you will want to know. First of all, any comet, even a Great Comet, is a faint and evanescent object. Consider that the

tail. In a shot taken with a sodium filter on April 16 at the Roque de los Muchachos Observatory on the Canary Islands, you can clearly see an appendage that does not resemble the two known tails. It is a completely gaseous tail, but composed of sodium in its neutral state rather than ions. It probably formed under the pressure of solar radiation, but because it was made of very light particles, unlike the dust tail, it tended to array itself in a straight line and with a directly antisolar orientation.

### Preparing for the Future

After comets Hyakutake and Hale-Bopp of 1996 and 1997, it seems unlikely that another such fortunate apparition will be repeated before the year 2000, but you never know. The table on page 29 clearly shows some very crowded visiting periods (for example, from 1858 to 1864, with four Great Comets in seven years; or 1880 to 1882, with three in three years) and some long, empty intervals (from 1702 to 1744 and from 1769 to 1811, for instance). In fact, our good fortune in the latter half of the 1990s simply made up for the barren 1980s, when there were no apparitions.

Naturally, I hope the reader will be able to share the great emotion of such a totally unpredictable event as soon as possible. Even after years of waiting, even after a promising comet has been discovered, any prediction is highly risky. In the end, it means keeping up with scientific—and especially astronomical—information in the years to come. This is not particu-

*Top right, wide-field image revealing a sodium tail discovered on Comet Hale-Bopp. (The sodium tail is the straight one on the left; the curved one at right is the dust tail.) The plasma tail is not visible in this image because a special filter was used. (G. Cremonese, A. Fitz Simmons and D. Pollacco.)*

*Bottom right, Hale-Bopp was highly visible from midnorthern latitudes until early in May. Giuseppe De Donà took this image on May 1. Note especially how the ion tail (which, at that point, was invisible to the naked eye) has been reduced to a few subtle and widely spaced filaments.*

density of a comet's tail is thousands of times lower than that of the Earth's atmosphere. It is, therefore, absolutely necessary to observe under the clearest sky possible, far from light and air pollution—better still, in the hills or mountains. The author, in fact, was able to fully enjoy observing Hyakutake and Hale-Bopp because he saw them under dark skies at altitudes almost always over 1,000 meters and often over 2,000. By comparison, people who observed from the open plains could see only one-third of the comets' tails. Even without climbing mountains, however, everyone ought to know how to get the most out of this extraordinary celestial event. In fact, if the atmosphere is clear enough, even someone living in a big city can take great pleasure in observing from modest altitudes and under moderately polluted skies.

The naked eye is the best instrument for observing a Great Comet. It is the only one that can contain the whole view. The eye, however, must be adapted to the dark, which is to say that you cannot expect to see a comet, no matter how prominent it is, right after you get out of your car or your lighted-up house. The pupil of the human eye reaches its maximum dilation of about seven millimeters only after about 20 minutes in the dark. Use this time to orient yourself if the place you have chosen for your observing session is relatively new or to find familiar constellations and discover new ones if you already know the sky. When you use a light to read, to make notes or to consult star charts, be sure to use a red one. It won't blast the human eye and cause you to lose your dark adaptation.

If the comet is spectacular but has a short tail, even small binoculars like opera glasses will boost the tail's visibility. Larger binoculars, such as the ones almost everyone has around the house, will help both to locate and to observe the comet in the unfortunate circumstance that it is fainter than predicted. In any case, it will be worth the effort to try to observe the object with all available instruments to confirm its individual characteristics and to gather novel details. If you have a spotting scope or telescope on hand, you should use the minimum magnification available in order to encompass as much of the comet in the field of view as possible.

Again, we must underline the great value and versatility of an instrument like binoculars, not only for comet observing but for astronomy in general. There's a reason why they are the chief tool of nonprofessional comet observers. People who have binoculars in the house and have never used them to look at the Moon or constellations should point them skyward. There is much to be enjoyed. Their limited magnification is largely compensated for by the ability to watch with both eyes and thus experience a depth of field, or three dimensionality. Compared with telescopic observations, binoculars offer the impression of truly being immersed in space.

Most binoculars are suitable for skywatching, but the best have an exit pupil closest to the size of the dilated human pupil. The exit pupil is the diameter of the cone of light coming out of the binocular eyepiece. It equals the ratio between the lens diameter in millimeters and the magnification. These two numbers are stamped on all binoculars. For example, a pair of 7x50 binoculars enlarges 7 times and has two 50-millimeter-diameter lenses. The exit pupil is 7.1 millimeters and is therefore optimal because it exactly equals the diameter of the human pupil. An 8x30 pair (30 ÷ 8 = 3.75), on the other hand, is not bright enough. In practice, you need an exit pupil of 5 or more for instruments with a small diameter (for example, 10x50, 8x40, etc.) and 4 for binoculars with a larger diameter (10x60, 15x70, 20x80, etc.). The most you will want for observing comets, whether exceptionally or moderately bright, is 20x80 binoculars, but their use will make a tripod absolutely necessary. We strongly advise using this accessory even with smaller binoculars, however; the added enjoyment of astronomical observing that results from the absolute stability of an image is priceless. Any common camera tripod is adequate. Naturally, its weight should be proportionate to the instrument being supported. Almost any camera shop or telescope dealer can furnish an adapter that will attach the binoculars to a tripod.

Moonlight can wash out faint comas and tails, so you are wise to forgo observing when there is a Moon in the sky, unless it is reduced to a thin crescent. Another important recommendation concerns the best time for observing. Frequently, a Great Comet is seen only quite low to the horizon in the morning or evening, so it really should be observed in the moments just at the beginning or end of the night. This is the only way to see the object at the highest altitude possible, in

*Top right, photo of Hale-Bopp taken on April 7 by Giuseppe De Donà (60-second exposure on Kodak Panther 1600 film processed at 3200, taken with a 50mm lens at f/1.8). Although this shot was made with the aid of a telescope's motorized clock drive, a camera fixed on a tripod would have achieved a similar result.*

*Bottom right, Hale-Bopp was so bright that it could even be seen from a balcony in the midst of urban streetlights, as this shot by the author testifies. The 5-second exposure was taken April 12 with a 55mm lens, fixed camera on tripod and 3200 ISO film.*

a perfectly dark sky. Remember that on average, astronomical twilight (signaling the beginning or the end of night) ends 1.5 hours after sundown and begins 1.5 hours before dawn. Thus it is best to be prepared. Habituate your eyes to darkness at least half an hour earlier both in the morning and in the evening so that you can readily locate the phenomena to be observed.

One last recommendation concerns clothing. It's cold at night, and not just in winter, especially when you don't move around for an hour or more and when you're in the mountains. It's a good idea to think ahead about the weather, particularly if you want to set up far away from home, because it will be impossible to remedy clothing that isn't warm enough. Always carry

winter clothes in the car. You will use them soon enough.

**How to Photograph a Great Comet**
The apparition of a Great Comet will be all the more memorable if you are able to photograph it. This is easier than it seems, and everyone can do it. A camera with a 50mm lens is adequate

*Above, the comet on March 29 from Duran Pass, with Mount Pelmo on the left. Photograph by Giuseppe De Donà using a tripod.*

63

when used at maximum aperture and focused on infinity, mounted on a tripod and furnished with a flexible cable release and very sensitive film. By keeping the shutter open (using the "B" setting), the cable release will permit exposures several seconds long. The film must be at least 1600 ISO. The fastest slide film (with an adequately suppressed grain) is Scotchchrome 800–3200 developed at 3200 ISO, while the fastest for color prints is Konica 3200. In their absence, you can opt for any 1600 ISO film or, at a minimum, 1000 ISO (about three times slower than 3200), which can eventually be "pushed," or processed at a higher rating (even 3200, although image definition will suffer).

Unfortunately, there is a limit on the exposure time beyond which the comet and any stars in the frame tend to move and produce bright streaks on the film due to the Earth's rotation. The maximum allowable exposure with the lens in question is 10 to 15 seconds, depending on the region of the sky in which the comet appears. At that setting, you obtain an image which is almost perfect, and the stars keep their point form. If it is bright enough, the tail will be seen in its entirety with 3200 ISO film. By tolerating a very slight or just barely perceptible movement with slower films, you can prolong the exposure to 20 or even 30 seconds. If your film is not very sensitive—from 400 to 1000 ISO—we still advise you to take longer exposures of a minute or even two if you have a chance to frame a particularly appealing foreground or landscape element, such as trees, mountains (maybe with snow cover and/or illuminated by a crescent Moon), a sea or lake or anything else that evokes a flight of fancy. In this case, the movement, even if it is clearly perceptible, will be perfectly tolerable, not only in an astronomical context but in an aesthetic one as well (in any event, the star trails will be very short). If the comet's tail is especially bright, like Hale-Bopp's, it will even be possible to use a medium telephoto lens and medium-length exposures—around 30 seconds, tops—to accomplish a greater magnification with limited movement. If the tail is very bright, it is always

Facing page, Hale-Bopp from the Nuvolau refuge (2,650 meters) in the Dolomite Alps, taken March 15 with a 35mm lens and fixed camera by Giuseppe Menardi of the Cortina Astronomical Association.

Top, Hale-Bopp photographed by Giuseppe De Donà on April 5 over Corvara with a 28mm lens and fixed camera.

Bottom, Hale-Bopp on April 1 above Pomagagnon, the mountain that dominates Cortina. Photograph by Alessandro Dimai and Renzo Volcan using a tripod-mounted camera with a 50mm lens and 35-second exposure.

possible to attempt a composition including a moderately illuminated city; in this case, exposure times have to be rather short. Artificial lights, alas, are a lot stronger than celestial ones.

Astronomical photographs are always enhanced if planets or bright stars, a thin lunar crescent or unusual constellations appear in the frame. Keep the long side of the frame parallel to the horizon for this type of composition. Everyone can experiment with composition, however, turning the camera one way and another to improve the overall image. Many of the best comet photos published in astronomy magazines contain a landscape element.

You should also note that if the tail of the comet is long enough, it is better to frame the head of the comet in a corner of the image to capture as much of the tail as possible (you cannot always see the whole tail with the unaided eye, so it sometimes gets cut off in the photograph). Do not frame the head too close to the edge, however, because a lens at widest aperture will register considerable distortions in that area, and that would compromise the

*Above left, this 30-second exposure of the comet was taken on April 4 with a 135mm lens by Gianvittore Delaito of the Feltre Astronomical Association Rheticus.*

*Above, Hale-Bopp photographed on April 18 by Giuseppe De Donà, using a tripod, at Vigo di Fassa. In the background, the Rosengarten.*

*Left, this photo by Corrado Marcolin, taken on March 29, shows how the brighter image obtained with an f/1.4 (50mm) lens permits the use of a less sensitive 400 ISO film. This, in turn, achieves less grain and greater resolution and keeps the exposure time to an acceptable 60 seconds.*

value of the photo. Try positioning the coma halfway between the center and the edge of the frame.

Find as dark a site as possible and operate in conditions of perfect darkness, with the comet at the highest altitude. This setup is capable of recording the complete length of a Great Comet's tail, including whatever is visible to the naked eye and more. With a 50mm lens at f/1.8 and Scotchchrome 800-3200 film, in fact, you can record stars much fainter than the ones visible to the unaided eye. In a 15-second exposure, you can record stars to magnitude 8.5. That is to say, you can capture stars 10 times fainter than those visible to the naked eye. Naturally, you get even better results if the available lens is a bright one that opens to f/1.4 instead of f/1.8.

It's a good idea to shoot an ordinary daytime picture at the beginning of a roll of film you later use to shoot a comet. This will facilitate the photo laboratory's work of threading the slides or cutting the negatives in strips without running the risk of cutting a frame in half. It may even be better to order the film left whole and uncut. If your camera has an LED display in the viewfinder, it's a good idea to take a few celestial test shots before shooting a comet. Shoot a few constellations, for example, to verify that the LED does not shed light on the frame of film. If you cannot extinguish the LED and if the camera's shutter is mechanical, remove the batteries—the shutter will still function. Otherwise, you may want to check with a photographer you trust.

*The author captured this striking image of the comet reflected in a mountain pond on April 1.*

# Meteor Showers

*The sight of a falling star, or meteor, is an experience that many people have never had.*

*There are times during the year, on a few special nights, when you can watch dozens or even hundreds of meteors fall.*

*On very special nights—one is expected before the end of this century—thousands of bright sparks appear to furrow the cold night air, and we feel the Earth sweep through the depths of space at a vertiginous speed.*

# Meteor Showers in History

**The Phenomenon of Falling Stars**

It's hard to believe, but many people have never seen a "falling star," or meteor. This is incredible because with the merest care and perseverance, one clear, dark night of observing will result in seeing not one but many, even dozens of, meteors on some special nights (special, but not that rare). For example, around August 11, observers under ideal conditions will catch the flash of 50 to 100 meteors an hour. This event is called the Perseid meteor shower, known to Europeans as "the Tears of St. Lawrence." And, as we shall see, there are at least two other periods of the year—January and December—that offer activity on the same order of magnitude. Rarer and much more remarkable, though, are so-called meteor storms, when as many as 10,000 to 20,000 meteors fall every hour. One such storm has been predicted to occur near the end of the century. These storms are a phenomenon that truly captures public attention. Their charm is enhanced by the fact that unlike the apparition of a comet, meteors have always been associated with something positive in the popular imagination—making a wish on a falling star, for instance.

Before looking at the history of meteor showers, it will be helpful to spend a few moments clearing up the terminology. To begin with, the terms falling star and shooting star are com-

On the preceding pages, a fireball photographed on August 13, 1994, at Cusercoli, Italy, by Stefano Moretti and colleagues of the Forlivesi I levelius Group using a 16mm fish-eye lens at f/2.8 and 800 ISO film.

Left, a fanciful representation of the 1833 Leonid shower over Niagara Falls as depicted in a 19th-century etching.

Above, a photograph of the Perseid shower taken at its peak on August 12, 1991, by Japanese astrophile Maroshi Hayashi. Twelve meteor trails are visible in the original, some of which are unfortunately lost in the printing process.

pletely inappropriate. The exact scientific term is meteor, which stands for the optical phenomenon produced when an object called a meteoroid burns up as it enters the Earth's atmosphere. Due to its small size, a typical meteor does not reach the Earth's surface. A larger object that does reach the ground is called a meteorite, like the one that recently plunged to earth in Italy on the outskirts of Fermo. While the objects that become meteors have a cometary origin, those which produce meteorites generally derive from asteroids.

When it enters the atmosphere, the object that produces a meteorite usually becomes an exceptionally bright meteor, called a fireball (or bolide). This latter term, however, is not officially defined (as compared with those officially sanctioned by the International Astronomical Union in 1961) and can be used to designate any meteor that is very bright, including those

produced by comet particles. Seeing a fireball is one of the most emotional and even disturbing moments that the starry sky can offer. For devoted skywatchers, it is not an exceptionally rare moment; the author has had occasion to observe at least 30 of them. The period from February to April seems to be more favorable for observing fireballs.

## Fire From the Sky

Meteors have been known, of course, since the dawn of time, from the moment humans turned their eyes to the sky. The first systematic observations date to about 2000 B.C. and were recorded in Korea and China. Sporadic mentions are found in various Chinese chronicles. For example, in 36 A.D., we find the first reference to the Perseid meteor shower: "More than 100 meteors were seen darting about in the morning hours." The first mention of the

Eta Aquarid meteors dates to 466: "Countless meteors, great and small, appeared toward the west." In records from 585, a reference to the Orionids is found: "Hundreds of meteors scattered in every direction"; in 687, the Lyrids: "Stars fell like rain from the sky"; and in 1002, the Leonids: "Lots of little stars fell."

The Leonids were also referred to in Arab chronicles. For example, in mid-October of 902: "An infinite number of stars was seen during the night, scattering like rain right and left, and that year came to be known as the year of the stars." And in 1202: "During the night of Saturday, in the last day of Muharram, the stars . . . flew at one other like a swarm of grasshoppers."

References to various events are found in historical European sources. The *Anglo-Saxon Chronicle* records that in 1095, "at Easter, on the night of Saint Ambrose and for most of the night, a giant host of stars seemed to fall from

*Above left, a drawing of a fireball from* Le ciel *by A. Guillemin.*

*Above right, the meteorite that fell at Fermo, Italy, on September 25, 1996. Its dimensions are 19 by 24 by 16 centimeters, and it weighs about 10 kilograms. (From* L'Astronomia*)*

the sky, so frequent that no one was able to count them." There is a reference to the Perseids' activity in 1243 in Matthew Paris's *Historia Anglorum*: "On July 26 of this year, on a perfectly clear night, brilliant stars were seen falling, scattering here and there in a manner so copious that if they had been real stars, none would be left in the sky."

In modern times, references to brilliant fireballs are fairly frequent, starting with the extremely famous one of November 7, 1492, when a huge meteorite weighing 50 kilograms fell near the Alsatian town of Ensisheim. It was observed and painted by Albrecht Dürer. In 1676, Italian astronomer Geminiano Montanari of Bologna observed a fireball whose appearance was followed by a sibilant, clattering noise. In 1707, in Worcestershire, England, one appeared which was so bright, "it made the stars and Moon disappear" and left traces, in the form of a small white cloud, that lasted for half an hour.

On March 17, 1719, Edmond Halley, the English scientist whose name is attached to the most famous comet, was astonished when a ball of fire shot out of the Pleiades toward Orion's belt. It was only a little less bright than the Sun, and it left a trail that lasted for more than a minute. In 1758, one seen at Newcastle, England, was so bright that its light "would allow a pin lost along the road to be found." Another fairly well-known fireball was observed on August 18, 1783, in both France and the British Isles (it was seen by the great astronomer William Herschel and poet William Blake). It was brilliant, it fragmented in air, and it was immortalized in a pair of paintings.

Among recent fireballs, we can cite one on August 10, 1972, that was almost as bright as the Sun. It was seen, photographed and filmed in the skies over Montana and southern Canada and was produced by an object a few meters in diameter that just missed crashing to earth by skipping off the outer atmosphere. Another one was seen shooting over the Emilia-Romagna region of Italy on January 19, 1993; its luminosity was halfway between that of the full Moon and the Sun.

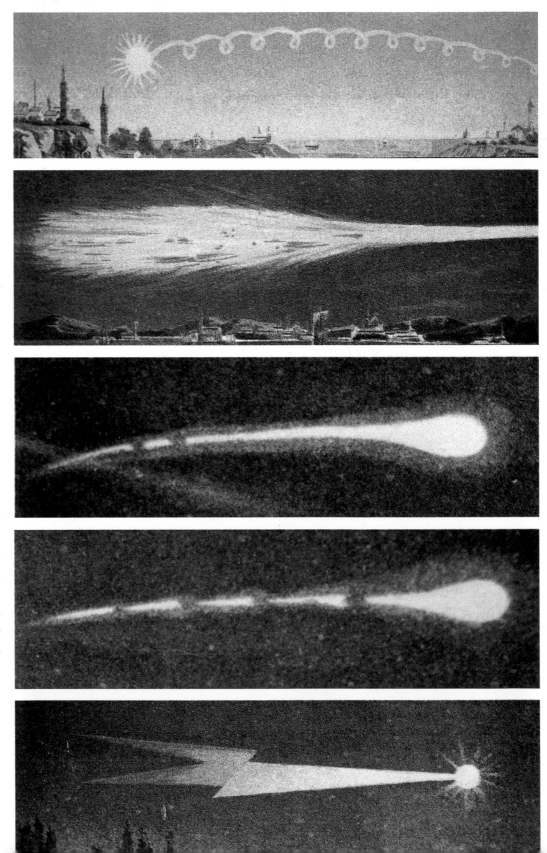

*A series of fireball illustrations taken from G. Naccari's* Atlante astronomico *(Astronomical Atlas). Some of these representations are really quite fantastic.*

Like comets, meteors were believed in ancient times to be purely atmospheric events. The word meteor derives from the Greek *meteoron* (literally "a thing up high, above the earth"), which is also the root of the word meteorological, used to describe phenomena that occur in the lower atmosphere. According to Aristotle, whose beliefs were widely held until the 18th century, meteors, like comets, were the result of hot exhalations from underground that caught fire through friction as they rose into the atmosphere.

In the 1700s, the idea that meteors might have an extraterrestrial origin began to take hold. The first to suggest this possibility, in 1714, was Edmond Halley. By comparing observations of various fireballs made at Oxford and Worcester in 1719, Halley was also able to calculate their altitude, which proved to be 119 kilometers. In 1794, Ernst Chladni, the father of acoustics, affirmed that meteors, meteorites and fireballs were of cosmic origin. He thought that the source of meteors and fireballs might be the ingredients left in interplanetary space after the formation of the planets. In 1798, two German students, H.W. Brandes and J.F. Benzenberg, repeated Halley's experiment using parallax reckoning to calculate the altitude of meteors. At first, their results were compromised by the small base dimension used for triangulation (10 to 15 kilometers), but by increasing the distance between observing sites, they found average values of 98 kilometers that are quite good even according to modern standards.

On November 12, 1799, thousands of sparks rained from the sky, casting Venezuela's inhabi-

73

*Above, German painter and engraver Albrecht Dürer witnessed the huge fireball of 1492. His painting, The Sphere of Fire, depicts the last phase of the apparition, probably the explosion of the meteoroid that generated it, a fragment of which fell near Ensisheim in Alsace.*

*Right, the biggest fragment of the meteorite that fell at Ensisheim is kept in the city: it weighs 54 kilograms and is 32 centimeters across. Photograph by Thomas Marvin.*

tants into a state of fear and confusion. The event was observed by the famous naturalist Alexander von Humboldt and French paleontologist Aimé Bonpland, who were together on a five-year expedition in South America. Bonpland reported so many meteors that "there was no area bigger than three times the Moon that was not constantly brimming with meteors." The shower was also seen in other parts of South America and off the coast of Florida. Aboard his ship at three in the morning, Andrew Ellicott was called to witness the phenomenon, and he wrote in the ship's log: "The whole sky seemed to light up with heavenly rockets that only disappeared when the Sun came up." The spectacle was also seen a little before dawn in the British Isles and even in Germany, despite daylight. Talking with South American natives, von Humboldt and Bonpland found evidence of another shower in 1766 and at various other times in the past, in a cycle of about 30 years. Von Humboldt realized—and was the first to do so—that the meteors all seemed to come from the same part of the sky.

In 1832 and 1833, there were two other spectacular meteor showers. In November 1832, observers in the Urals, Arabia, Mauritius, the North Atlantic and various parts of Europe saw an impressive shower. But the following year, some real fireworks were seen in the eastern regions of North America. Tens of thousands of meteors fell in the early hours of November 13, 1833. Understandably, bystanders were once again awestruck. People ran to wake up their neighbors so that they could witness the scene, but many woke up suddenly on their own due to the alarming blaze of the brightest fireballs lighting up their dark bedrooms.

*Top, English painter Thomas Sandby was among those who saw the fireball of August 18, 1783, from the terrace of Windsor Castle near London. This is one of his watercolors depicting the event.*

*Bottom, the August 10, 1972, fireball photographed over the Grand Tetons in Wyoming. Photograph by James Baker.*

A certain Professor Thomson of Nashville, Tennessee, who was yanked out of bed for the phenomenon, wrote that it was "the most sublime vision I have ever had." A planter in South Carolina awoke to the desperate cries of slaves on his and two nearby plantations. He heard a voice calling him to come out and yelling, "Oh, my God, the world is in flames!" Dashing outside, he did not know whether to be more amazed by the celestial performance or by the crowd facedown on the ground imploring God to save them and the world. Another eyewitness, Strickland, reported that "two fireballs [were] half as bright as the Moon" and that "several hundred of these gorgeous stars were visible at once . . . leaving a long streak of flame in their tracks."

*Above, an etching showing the great Leonid storm of November 12, 1799.*

A few witnesses reported that the meteors "fell as thick as snowflakes." In Boston, a number of meteors equivalent to half the number of snowflakes in a storm was estimated. A slightly less poetic and more quantitative estimate spoke of something like 240,000 meteors during the incredible nine hours the event continued, essentially the entire night. In New York, a rate of 10,000 falls per hour was estimated, with some meteors as bright as the full Moon. Historian R.M. Devens declared the meteor storm of 1833 one of the 100 most memorable events in the entire history of the United States. A number of scientists, including Denison Olmsted, professor of physics at Yale College, observed the phenomenon as well. Such observations confirmed von Humboldt's impression that the meteors all came from the same point in the sky, in the constellation Leo (hence the name Leonids). It was also confirmed that the Leonids moved through the sky along with the constellation, participating in the sky's apparent diurnal rotation.

The explanation is an effect of perspective: When the Earth crosses a stream of particles in space—the trajectories of which are actually parallel to each other—they appear to diverge from one point as they approach us from our viewing position. This is the same effect as, for example, the sides of a street that seem to diverge from a single point in the distance on the horizon. The point from which meteors appear to diverge is called the radiant.

Eventually, the hypothesis was advanced that shower meteors originate from comets with a very short period. The activity of a minor shower (from a few dozen to a few hundred meteors in the last hours of the night) had long been observed to repeat at irregular intervals around August 10 (the first record was made in China

*Right, the most famous image of the 1833 Leonids, a xylograph entitled* Meteor Shower of November 13, 1833. *It was made by Karl Jauslin and Fritz Völlmy, two Swiss engravers born after the event, and was published* *in Michigan in an 1889 book of biblical readings as an example of Old Testament prophecies being fulfilled.*

in 36 A.D.). Up to 1833, this had been noted about 15 times. In England and Germany, beginning in the first half of the 19th century, these meteors were called the "Tears of St. Lawrence," named for the saint whose feast day is August 10, the date of his martyrdom. In 258 A.D., under Roman Emperor Valerian, Lawrence was put to death by burning. According to the legend, the meteors were his flaming tears. In 1834, American physicist John Locke established that like the Leonids, these meteors also came from a single point in the sky, in the constellation Perseus. Between 1836 and 1837, Belgium's Adolphe Quetelet and American Edward Herrick confirmed that the shower had an annual rhythm. Among his several hypotheses, Herrick was the first to propose a cometary origin for meteors and to notice the activity of the Geminids and the Lyrids. And in 1839, Quetelet published the first systematic catalog of meteor showers observed since the Middle Ages.

## The Curtain Is Raised

In 1861, American astronomer Daniel Kirkwood suggested the currently accepted explanation for the origin of meteor showers—the

residue of particles of material (now known as meteoroid streams) from old comets, distributed along their orbits.

Between 1863 and 1865, American mathematician Hubert Anson Newton tried to calculate the orbits of two known meteoroid streams and discovered that they resembled comets' orbits. He also found an annual cycle for the Leonids and the Lyrids, which appeared in April. Completing his historical investigation, moreover, he noted that since the year 902, the most spectacular Leonid showers had been observed to occur at intervals of 33 or 34 years— in 902, 934, 967, 1037, 1202 and 1366. There had also been mentions of less spectacular Leonid showers in 931, 1002, 1101, 1533, 1602 and 1698. It was evident, then, that the distribution of the particles was not constant along their orbit; instead, they were clustered most heavily in a section that the Earth periodically passed through in a specific interval of time (33 to 34 years). Thus Newton was able to predict that the next great storm would be in November 1866. (Unbeknownst to Newton, German astronomer H.W.M. Olbers had reached the same conclusion a few years earlier.)

And there was indeed a storm on November 13, 1866. Though it was less profuse than those of 1799 and 1833, it was still a gorgeous show, and this time, Europe was favored with the view. Among the most vivid impressions was an account by English astrophile Robert Ball: "The meteors distinguished themselves not only by their great multitude but also for their intrinsic magnificence. I will never forget that night. That evening, I was occupied in my usual observation of nebulas with Lord Rosse's large reflector. Naturally, I knew that a meteor shower had been predicted, but nothing I had heard prepared me

*Top, when meteors rain down, they seem to emerge from a single point in the sky called the radiant.*

*Bottom, Lord Rosse's 183-centimeter-diameter telescope, built in 1845, was by far the world's largest in its time. (Science Museum, London)*

for the splendid spectacle that we would see. It was almost 10:00 when my assistant's yell made me detach from the telescope just in time to see a beautiful meteor cross the sky. It was quickly followed by another and then by others in groups of two or three . . . at the last, Count Rosse (then Lord Oxmantown) met me at the telescope [and] we decided . . . to scale the summit of the wall of the great telescope where the view opened onto the whole sky. There, for the two or three hours that followed, we witnessed a spectacle that will never be canceled from my memory as long as I live. The falling stars gradually grew in number until a lot of them were seen at once . . . but they all came from the east. As the night went on, Leo rose above the horizon, and then the characteristic salient of the shower became obvious. All the trails radiated from there."

Ball said that he could not estimate the number of meteors he saw, "every one of which would have been bright enough to provoke a note of awe on any ordinary night," but 9,000 were counted at the Royal Greenwich Observatory. The high point came between 1:00 and 1:30 a.m., with more than 120 sightings a minute. Another count, taken at Twickenham by Hind and three assistants from 1:00 to 1:07:05, totaled 514 meteors. When the peak arrived at 1:10, it became impossible to count them.

During the shower, British astronomer John Herschel focused on the appearance of four extremely bright meteors whose trails persisted from 5 to 14 minutes, leaving behind balls of light that drifted with the high-altitude winds for 8 to 10 degrees before they dissolved. The last, in particular, "produced a majestic flaming ring 3 or 4 degrees wide. It was visible for 14 minutes, gradually expanding into a heart-shaped noose until it included the principal stars of Ursa Major

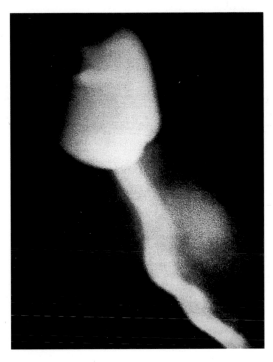

in a wide, fantastic corona before it disappeared." Even the meteors' colors claimed attention; at Greenwich, the majority were described as blue, the remainder divided among white, blue-white, yellow, red and green. Herschel reported that most were white, but a few were orange or emerald green. The following year, the event repeated, though on a comparatively minor note. One observer in North America counted 1,000 falls per hour, despite the presence of the nearly full Moon. The event repeated again in 1868 with about the same intensity.

In 1866, Italian astronomer Giovanni Virginio Schiaparelli came on the scene, perhaps inspired by his direct observation of that year's Leonid shower (he also observed the Andromedids in 1872). He established a definitive connection between comets and meteors and was the first to connect a particular shower to a par-

ticular comet. Schiaparelli determined that the parent comet of the Tears of St. Lawrence was Swift-Tuttle, which had been discovered by two Americans four years earlier. He also proposed the name "Perseids," after the constellation in which the radiant is located.

During this same period, Italian astronomer Angelo Secchi (one of the great pioneers of stellar spectroscopy) took new meteor parallax measurements using as a base the distance between Rome and Civitavecchia (65 kilometers) and finding altitudes of 75 to 250 kilometers. In 1867, Schiaparelli and, independently, Austrian astronomer Theodor von Oppolzer and Germany's Carl Peters identified the parent comet of the Leonids as Tempel Tuttle, discovered just a year earlier. Still in 1867, Austrian astron-

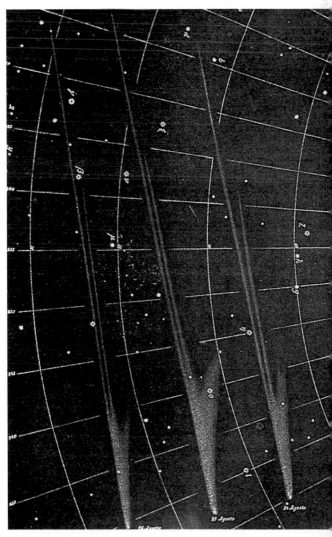

*Above, an extremely bright fireball that appeared in the United States in 1933 left a lasting trail and a ball of light which lingered in the sky for more than 1.5 hours. Photograph by C.M. Brown.*

*Right, the 1862 apparition of Comet Swift-Tuttle, parent of the Perseids, became quite prominent. Here, it is shown in a few sketches made during August—the period of peak visibility— by astronomer and famous*

*comet hunter Ernst Wilhelm Tempel.*

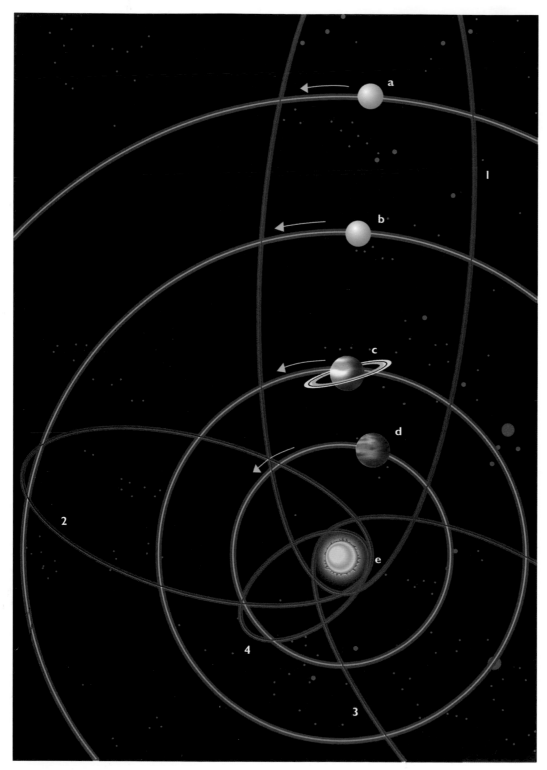

*Seen from above the plane of the Earth's orbit are the orbits of the meteoroid streams producing the Perseids (1), the Leonids (2), the Lyrids (3) and the Andromedids (4)— and of their respective parent comets Swift-Tuttle, Tempel-*

*Tuttle, Thatcher and von Biela. Note, however, that despite the appearance of this two-dimensional drawing, the comet orbits pictured actually lie at a significant incline to the Earth's orbital plane.*

a. Neptune
b. Uranus
c. Saturn
d. Jupiter
e. Sun

omer Edmund Weiss found two other relationships between meteors and comets: the Lyrids were connected with Comet Thatcher, discovered in 1861, and the Andromedids (typically a November stream) with Comet von Biela, with a cycle of about seven years (hence they are sometimes called Bielids). Weiss even ventured to predict that a truly spectacular Bielid shower would occur around November 28 of either 1872 or 1879. On November 27, 1872, an impressive display took place. In the evening hours between 5:30 and 11:50, an observer at Greenwich counted at least 10,000 meteors. But the fate of Comet von Biela itself is so curious that it merits further discussion.

**The Strange Case of Comet von Biela**

Jacques Leibax Montaigne, a French pharmacist and amateur astronomer, discovered Comet von Biela in 1772. No one realized that it was periodic until 1805, when Jean Louis Pons observed it again and, by calculating its orbit, Friedrich Wilhelm Bessel identified it with the one seen more than 30 years earlier. The comet's third visit was observed in 1826 by the man who was eventually honored with having it named after him, Austrian infantry captain and noted amateur astronomer Wilhelm von Biela. The comet's period was calculated at 6.7 years, and its predicted return happened right on schedule, in 1832. On its next visit, in 1839, it was poorly positioned for observation, but in 1845–46, it was seen again, along with a surprising phenomenon.

On December 29, 1845, a fainter comet showed up alongside the main one. The two objects were at first linked by a bridge of light but gradually separated until they displayed two comas and two distinct parallel tails. At a certain point, the secondary comet became

brighter than the primary one. By the time they faded from view, the two fragments were separated by about 300,000 kilometers. In 1852, Comet von Biela appeared for the last time. Angelo Secchi sighted it at a remarkable distance away from the predicted position, and it still appeared to be a double object. The two pieces were now farther apart, however, separated by at least two million kilometers. Once again, the secondary comet was at first dimmer, then it surpassed the main one in brightness. It was known that the comet was in a bad position for observation in 1859, but it was expected to become visible in 1866. That did not happen, however, and no trace of Comet von Biela was ever found in the sky again. Those fortunate enough to observe the comet

in 1846 and 1852 had actually witnessed the death of a comet directly and for the first time.

After the 1872 shower, the Bielids produced an even more spectacular display in 1885, still coinciding with the now extinct parent comet's passage to perihelion. On the night of November 27, some British observers estimated counts of 75,000 meteors an hour.

## The Big Disappointment

Success in forecasting the Leonid storm in 1866 and the Bielids in 1872 led astronomers to believe that they now knew everything about meteor showers. To be sure, this was the epoch of Positivism and of the great triumphs of celestial mechanics, including the discovery, through calculus, of the planet Neptune and discoveries of innumerable multiple-star systems, which seemed to confirm the universal validity of Newtonian laws. The entire machine of Nature seemed to move according to established and rigid laws. Thus astronomers expected to see a great show when the Leonids returned at the end of the century, in November 1899. Two astronomers in England, Stoney and Downing, set to work calculating the exact conditions for the expected storm using the stream's known orbit, which had been determined about 30 years earlier by John Couch Adams, one of two astronomers to posit the existence of Neptune on the basis of perturbations in the orbit of Uranus. Taking into account perturbations by all the principal planets, it was found that the

storm ought to occur at 6:00 in the morning (Greenwich time) on November 15, 1899.

In 1898, the Leonids showed an appreciable increase in their rate of activity—50 to 100 an hour—and this seemed to herald the impending spectacle. An immense anticipation was created that also engaged the general public, who hoped to witness something as memorable as what had already happened twice in that century. However, when Stoney reviewed his calculations, he realized that the orbit of the cometary particles had been altered by close encounters with Saturn in 1870 and Jupiter in 1898 and that unless the stream was larger than anticipated, it would miss its meeting with the Earth. In fact, the meteoroids had been displaced by at least 1.75 million kilometers. On November 10, Stoney informed the Royal Astronomical Society, but it was too late. The shower activity that year attained a pitiful 40 meteors an hour, and thousands of people who had eagerly anticipated the event were hugely disappointed. As meteor expert Charles Oliver wrote in 1925, "The failure of the predicted Leonid visit in 1899 was the worst blow astronomy ever suffered in public esteem." Unfortunately, because of this disappointment, most people lost interest in the Leonids, and thus they missed a date with two remarkable succeeding showers—nearly 1,000 meteors an hour in 1900 and 2,000 an hour in 1901—that were seen by only a very few people in Canada and the United States.

*Above, a contemporaneous print showing the spectacular Bielid shower of November 27, 1872.*

*Right, Comet von Biela "doubled" in a drawing by astronomer Otto Struve on February 19, 1846.*

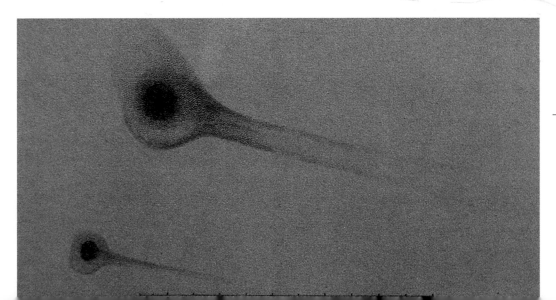

# The Origin and Nature of Meteors

## Comet Dust

The successful prediction of meteor storms for 1866 and 1872 linked to comets Tempel-Tuttle and von Biela undoubtedly bolstered astronomers' confidence in their understanding of the connections relating meteor showers, ordinary (or sporadic) meteors and comets. In the light of modern knowledge, this understanding is even more complete.

The particles that are blasted away from a comet's nucleus every time it swings close to the Sun become part of the dust tail. They eventually spread all along the comet's orbit. When the Earth, in its annual journey around the Sun, crosses that orbit or at least passes nearby (if the meteoroid stream is wide enough), there will be a fairly profuse meteor shower.

Virtually all the dust that is the source of meteors comes from periodic comets. Meteors originating from "new" comets—those which come directly from the Oort cloud or the Kuiper belt and are passing close to the Sun for the first time—are of an absolutely insignificant number. Moreover, short-period comets contribute far more to the creation of the interplanetary dust store than those with long orbits, because they visit our star more often, exposing their nuclei at length to solar-radiation pressure.

The dust streams can be older or younger and, therefore, more or less dispersed along the comet's orbit. For example, the Leonid dust is apparently still concentrated in a relatively restricted zone (primarily around the comet itself) of Tempel-Tuttle's orbit, which the Earth crosses about every 33 years, near the comet's perihelion passage. On this date, the showers are truly spectacular, while their activity in other years is almost negligible (10 to 15 meteors an hour).

The Perseid stream is obviously much more dispersed along the entire orbit of the parent comet, because although no Perseid storms are ever seen, a few dozen to a few hundred meteors are observed every year. The greater dispersion of the Perseids is also indicated by the fact that it takes the Earth a few weeks to cross this

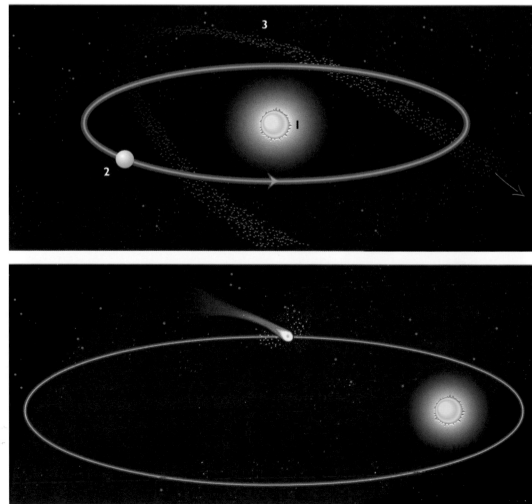

*Top, distribution of a meteoroid stream (3) along a comet's orbit. When the Earth (2) crosses such a stream, a meteor shower occurs. The Sun (1) is at center.*

*Right, the dust released by Comet Tempel-Tuttle is still fairly concentrated around the parent comet, which explains why the Leonid meteor storms occur at such regular intervals.*

stream (in other words, it is active for at least two weeks every year) but only one or two days to cross the Leonid stream.

The most dispersed streams are obviously those which feed the continuous backdrop of sporadic meteors that are seen throughout the year. Various factors cause this dispersion, including collisions among the particles, a loss of energy due to their interaction with solar radiation and the planets' gravitational influences. It has been calculated that meteoroids take from 1,000 to 10,000 years to distribute themselves uniformly along the orbit of their parent comet. The dispersed dust particles form a lens-shaped disk, centered on the plane of the Earth's orbit around the Sun, that extends all the way to Jupiter. This disk is sometimes visible just after sundown or just before daybreak when the sky is very clear. It takes the form of a luminescent pyramid rising from the horizon and is called the zodiacal light. The name comes from the disk's being visible against a backdrop of the constellations of the zodiac, corresponding to the orbital plane of the solar system. In midnorthern latitudes, this phenomenon is best seen in the evening around the vernal equinox or in the morning around the autumnal equinox. As it revolves around the Sun, the Earth crosses the zodiacal disk of dust like a ship plowing through ocean waves. Consequently, most sporadic meteors seem to come from the same direction—the point in the sky toward which the Earth appears to be moving. This point, called the apex, naturally changes over the course of a year but is always at a right angle with respect to the Sun's posi-

*The zodiacal light photographed by the author. Note the brief trail of a meteor that appears at right.*

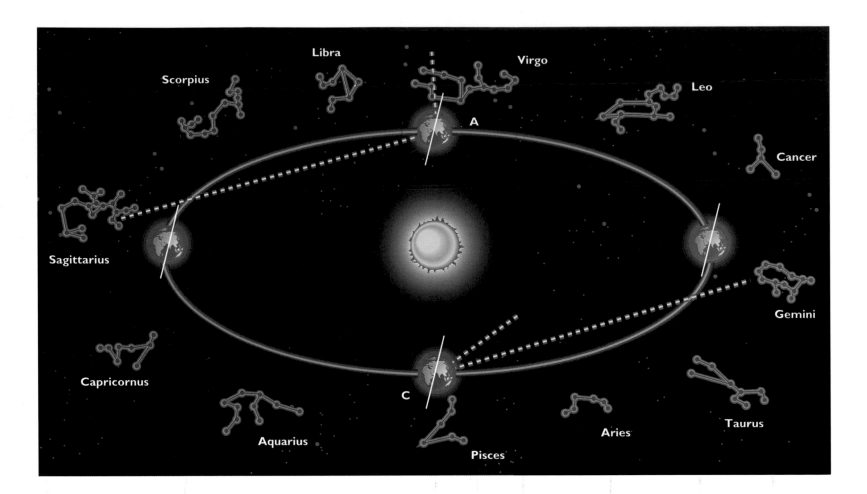

tion. From our latitudes, therefore, the apex sits very high in the sky in autumn, in the constellation Gemini, where the Sun is found in summer. It lies very low in the spring, however, in Sagittarius, at a point occupied by the Sun in winter. Given that any celestial object is better seen when it is high in the sky, because its light travels through less atmosphere and less smog, far more meteors will be seen in autumn than in spring from midnorthern latitudes. While 1 to 4 meteors can be seen every hour during the spring, 5 to 10 can be seen in autumn.

## Comet Breakups

The Bielids are naturally quite different from the Perseids and the Leonids, considering that their parent comet completely disintegrated after splitting in two. According to recent calculations by Brian Marsden of Harvard's Smithsonian Astrophysical Observatory and by Zdenek Sekanina of the Jet Propulsion Laboratory in Pasadena, this breakup occurred between 1842 and 1843. That the comet literally disintegrated is proved by the fact that otherwise, it—or whatever was left of it—would have

been seen in 1971, when it was due to pass Earth at only 7.5 million kilometers. To date, not one trace of this comet has been located.

That the Bielids produced only one other big storm after 1872 is amazing. In 1892, 1899 and 1904, there were minor events, then nothing. It has been hypothesized that only the comet's secondary nucleus was destroyed, while the primary one was violently perturbed by the forces of the breakup. According to this theory, the stream produced by the disintegration of the secondary nucleus passed very close to

*Due to the orientation of the Earth's axis, more sporadic meteors are seen in autumn than in spring from midnorthern latitudes, where the majority of skywatchers live. As the above figure shows, there is a much wider angle between the direction of the zenith and the Earth's direction of forward (counterclockwise) orbital motion at the vernal equinox (A) than at the autumnal equinox (C).*

Jupiter in 1901, where it suffered powerful gravitational perturbations, tearing it right out of the orbit that used to intersect the Earth's.

English astronomer Ivan Williams adds other details to this curious story. He determined that in 1832, Comet von Biela would have collided with the meteoroid stream from which the Leonid shower originates. It can't be proved that this destroyed its nucleus; other events may have caused the fragmentation, perhaps reinforced by the crossing of the Leonid stream. Recent computer simulations by David Hughes of Sheffield University indicate that the Bielids (aka Andromedids) could revisit our skies around 2120.

This was not the only time a comet has split or disintegrated. According to Sekanina, no

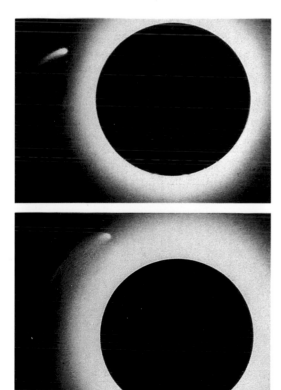

fewer than 20 other comets have met this fate. In addition to the cases described earlier—the Great September Comet (1882), Ikeya-Seki (1965) and West (1976)—there were breakups in 1860 (Comet Liaia), 1888 (Sawerthal), 1889 (Davidson), 1896 (Giacobini), 1899 (Swift), 1905 (Kopff), 1914 (Campbell), 1915 (Mellish), 1916 (Taylor), 1943 (Whipple-Fedtke), 1947 (Southern Comet), 1955 (Jhonda), 1957 (Wirtanen), 1968 (Wild), 1969 (Tago-Sato-Kosaka) and 1970 (Kohoutek).

There can be several causes for a comet's nucleus to break up. When it occurs in a Sun-grazing comet, like the Comet of 1882 or Ikeya-Seki, the prevailing cause is tidal action. The segments closest and farthest from the Sun undergo competing tidal forces, and due to the nuclear components' scant cohesion, this may result in disintegration. Incidentally, a massive planet like Jupiter can also induce these effects, but far less often than the Sun. Only two examples related to Jupiter are known, the first being Comet Brooks 2 in 1889. This object had already experienced several close encounters with the planet and finally ventured less than 145,000 kilometers from Jupiter—well inside the so-called Roche limit, within which no solid object can exist, because tidal forces pull it apart and shatter what's left. The second, now famous case is Comet Shoemaker-Levy 9 that was widely publicized in 1994 (see Chapter 2). The comet came within 21,000 kilometers of Jupiter, a close encounter that not only split the comet into 21 fragments but eventually caused them to tumble one after another into Jupiter's gaseous atmosphere.

A variety of explosive breakups can affect Sun-grazing comets, due to superheating. For

comets like West, whose perihelion is a little farther from the Sun, this solar action has a more indirect role related to a violent release of gas by the comet's nucleus once it heats up. In 1966, Whipple and Stefanik suggested a splitting mechanism for comets that break up farther from the Sun. As the comet flies through interplanetary space, it develops a crust of highly volatile elements on the surface of its nucleus due to heat produced internally by the decay of radioactive isotopes. Such a structure may break down when subjected to solar radiation, because it is not a very effective shield against the blast of heat and the differential expansion associated with it (caused by the unequal distribution of solar radiation hitting the nucleus). This structure would then break away from the rest of the surface, which could be the

*Above left, two stages of Comet Ikeya-Seki's close pass by the Sun in 1965. The comet paid for its gamble with the fragmentation of its nucleus. (Tokyo Astronomical Observatory)*

*Above, breakup of a comet's nucleus due to tidal forces experienced as it passes close to the Sun or a large planet.*

beginning of the end, because it would weaken the comet's cohesion.

Another mechanism suggested for the more distant comets is a change in the ices in the nucleus from an amorphous to a crystalline state when heated above –120 degrees C (the temperature that prevails in the space between Mars and Jupiter where most asteroids sail). In this process, the density of the ice goes from two to one gram per cubic centimeter, and a lot of heat is released.

In 1982, Sekanina offered another hypothesis that would apply to periodic and older comets. In the course of a comet's life, large areas of its nucleus would be coated with a thick layer of dust. This coat would finally weigh enough to modify the distribution of mass and, therefore, the rotation of the nucleus as well. The tensions created within this rapidly rotating nonspherical object would sooner or later defeat cohesive forces along any lines of structural weakness that might exist in the nonhomogeneous mix of ices and dust. Eventually, this entire fritterlike coating, along with a layer of underlying ice, would break away from the nucleus.

In effect, the splitting phenomenon seems, in most cases, to be due initially to smaller fragments detaching from the nucleus. They would break away precisely when the ice that holds the nucleus together melts under the onslaught of solar heat. Deprived of these fragments, the formerly smooth and regular nucleus assumes a rough, irregular shape. These newly created "angles" are more vulnerable to forces produced by the pressures of radiation and the solar wind; the mechanical cohesion of the nucleus is finally overwhelmed, and following additional melting of the ices, it breaks up.

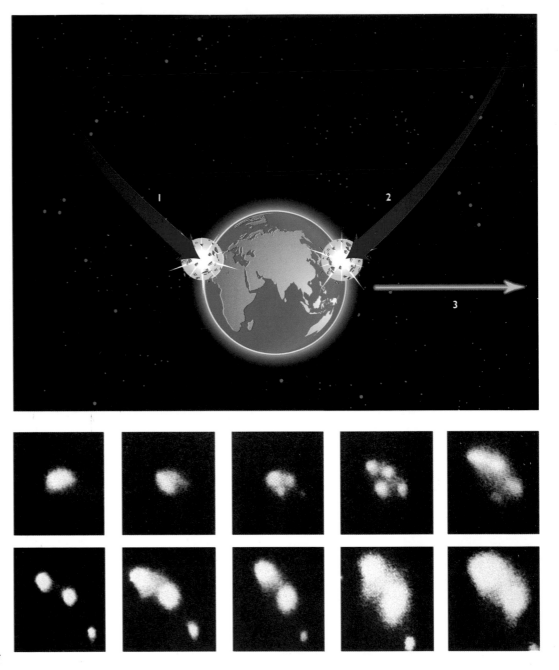

### Meteors Laid Bare

The meteoroid that produces a meteor is, therefore, a fragment of comet dust. Its modest dimensions vary from the size of a grain of sand to that of a pebble. A typical second-magnitude meteor is produced by a particle 0.1 gram in mass and one centimeter in diameter. The density of these particles is very low, typically only

*As shown in the drawing at top, the geocentric velocity of meteoroids depends on the geometry of their encounter with our planet in space.*
*1. Rear impact*
*2. Frontal impact*
*3. Direction of Earth's orbital motion*

*Above sequence of images shows the progressive fragmentation of Comet West's nucleus into four pieces on its 1976 passage. (New Mexico State University Observatory, Las Cruces)*

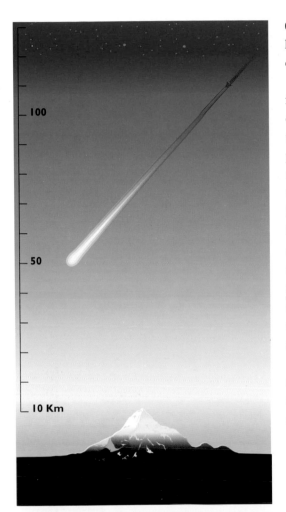

0.3 gram per cubic centimeter. Their structure has been compared to coffee grounds or grains of sugar.

The meteoroids' compactness varies widely from stream to stream. The Draconids, for example, are very flimsy, because they were recently released from their parent comet, periodic Comet Giacobini-Zinner, and have not yet undergone complete fragmentation. On the other hand, the Taurids, which are linked to Comet Encke, are much more resilient. At just 3.3 years, Encke has the shortest comet period, and because it has made so many trips around the Sun, it long ago released most of its dust, which has completely fragmented. The chemical composition of meteoroids is typical of comet dust, consisting of silicates with traces of heavier elements.

Meteoric dust collides with our planet at speeds ranging from 11 to 72 kilometers per second. We speak of geocentric velocity as the speed of dust perceived by someone standing on Earth, as though the planet were at a standstill. Of course, this is not the reality. In fact, the Earth moves around the Sun at a velocity of 30 kilometers per second, while the dust travels at about 40 kilometers per second, which is close to the 42-kilometer-per-second speed necessary for a particle of negligible mass at the distance of the Earth's orbit to escape the solar system. If the meteoroid approaches Earth from behind (when the meteoroid is moving in the same direction as Earth), the velocities of Earth and the meteoroid will be subtractive, and the object will fall with a speed just equivalent to the Earth's escape velocity (the minimum speed an object has at the moment it hits ground, 11.2 kilometers per second). If, instead, the Earth and a meteoroid crash head-on (when the meteoroid is moving in the opposite direction to the Earth's), the velocities are additive. In every other case, the speeds are somewhere in the middle.

Top, median heights at which meteors are first and last seen.

Bottom, a typical meteor produced by a small meteoroid. Photograph by Roberto Gorelli, UAI Meteor Division.

85

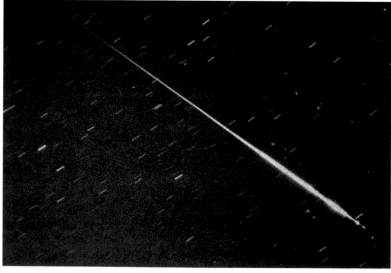

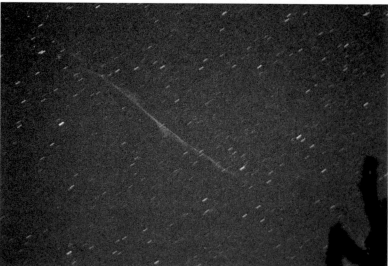

Top, the wider, brighter trail left by a fireball results from a meteoroid that is larger and/or faster than usual. Photograph by Roberto Haver, UAI Meteor Division.

Bottom left, an exceptionally bright meteor leaves a lasting trail (bottom right) in the sky. (Greeley Astronomical Society, Colorado)

## Passing Through the Atmosphere

At an altitude of somewhere between 80 and 120 kilometers, friction with the layers of the Earth's atmosphere renders the surface of a meteoroid incandescent and causes it to evaporate. The gases produced during evaporation collide with atoms of atmospheric gas and excite them, causing their electrons to jump to higher energy levels. When the electrons return to their initial levels, a burst of light is emitted, which causes the luminous trail we call a meteor. The trail is typically 20 to 30 kilometers long and a few meters wide. If the meeting between the meteoroid and atmospheric atoms occurs at higher speeds, and therefore at higher kinetic energies, the degree of ionization will be greater, and a greater quantity of light will be emitted. The resulting meteor will be brighter. At even higher speeds, the atmospheric atoms may be ionized, meaning that one or more of their electrons will be stripped away. When they recombine (when the atoms recapture electrons that were released into space), the result is an exceptionally bright meteor and one of the lasting trails that often accompany the brightest meteors. The meteor event, then, is very short. It typically lasts for no more than a few tenths of a second. Slow meteors lasting for one or two seconds are very rare. Sometimes a meteor seems to explode, or fragment, in the sky. Its trail continues after the breakup, even if it is faint, and sometimes there is more than one trail. In effect, the meteoroid is abruptly shattered by the intense heating of its surface combined with the enormous pressure exerted by the atmosphere.

The altitude at which the meteor disappears is variable and depends on its mass, its density

*This fireball appeared on the night of August 11/12, 1993—a Perseid as bright as the full Moon! Photograph by Severino Lodi, UAI Meteor Division.*

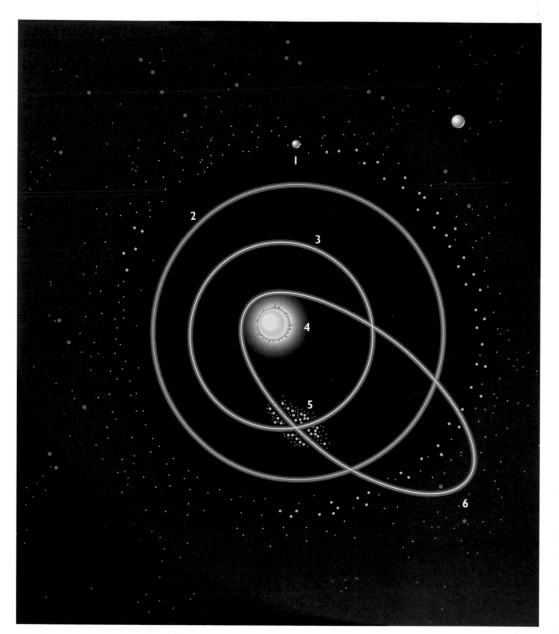

and the angle of fall. Generally, it is between 40 and 50 kilometers. At this altitude, even the biggest and densest meteoroids disintegrate completely, and no confirmed meteorites have ever been found following a meteor shower. It stands to reason that the dust fragments cannot be all that big; at most, they might be as big as a small but flimsy pebble. The appearance of an extremely bright fireball during a meteor shower is very rare. The brightest fireballs usually occur when a fragment of a small asteroid enters the atmosphere. The fireball observed during the Perseid shower on August 12, 1993, which looked as bright as the full Moon, was obviously an exception. The meteoroid that gave rise to it probably measured about 30 centimeters across, with a mass of 10 to 15 kilograms.

The total number of meteoroids entering the Earth's atmosphere each day is vast—perhaps 25 million. Their total mass, however, is less than one ton. In addition to these, no fewer than 400 tons of micrometeorites, having dimensions on the order of a micron or less, fall onto our planet from outer space each day. The atmosphere brakes these particles before they are able to incandesce, and they are slowly deposited on the ground within a few days or weeks.

### The Principal Meteor Showers

Thus far, we have referred to only a few meteor streams, but at least several dozen are known. The Meteor Division of the Unione Astrofili Italiani (UAI) records at least 100, for example, in its 1997 *Almanacco Astronomico* (*Astronomical Almanac*). According to David Hughes,

*Above, the orbit of the asteroid Phaethon and the meteoroid swarm associated with the Geminids.*

*1. Asteroid belt*
*2. Orbit of Mars*
*3. Orbit of Earth*
*4. Sun*
*5. Geminid swarm*
*6. Orbit of Phaethon*

*Left, the major meteor showers and their principal characteristics.*

### MAJOR METEOR SHOWERS

| Name | Peak | Duration | ZHR | Speed | Parent |
|---|---|---|---|---|---|
| Quadrantids | January 3 | 0.4 | 130 | 42 | |
| Lyrids | April 22 | 1 | 15 | 48 | Comet Thatcher |
| Eta Aquarids | May 6 | 6 | 55 | 66 | Halley's Comet |
| North Delta Aquarids | August 12 | 8 | 12 | 41 | |
| Perseids | August 12 | 3 | 90 | 60 | Comet Swift-Tuttle |
| Orionids | October 17 | 2 | 30 | 66 | Halley's Comet |
| Leonids | November 18 | 2 | 45 | 72 | Comet Tempel-Tuttle |
| Geminids | December 14 | 3 | 105 | 36 | Asteroid Phaethon |
| Ursids | December 22 | 1 | 40 | 34 | Comet Tuttle |

there are nearly 200 streams, but the majority have a zenithal hourly rate (ZHR) of three or less, and they remain unknown. The ZHR is how many meteors a single observer sees in one hour, assuming ideal observing conditions—that the sky is completely clear and free of natural or artificial obstructions and that the radiant is at the zenith. (In poor observing conditions, the actual hourly rate observed can be 10 or 20 times lower than the ZHR.)

There are a dozen major meteor showers whose ZHR exceeds a rate of 10. The activity of the others is low enough to be considered doubtful at times, in the sense that it can be confused with the continuous background formed by the sporadic meteors. The table on page 88 lists the nine principal showers, excluding meteors that fall during the day (detectable only by radio signals). Listed are

the date of the shower's peak, the duration in days of the most intense activity, the median ZHR over the past few years (from the Meteor Division of the UAI), the average speed of the meteors in kilometers per second and the parent comet, if known. As noted previously, the name of a meteor shower derives from the constellation in which its radiant is located. The Quadrantids were named after a constellation called Quadrans Muralis, which became obsolete in 1922; their radiant lies in what is now part of the constellation Bootes.

Regarding the velocity, it can be said that the higher it is, the higher the altitude at which disintegration of the meteoroid begins. The trails

89

*Left, a close-up of Halley's nucleus produced from a composite of six separate images taken by the Giotto probe. The darker regions on the right are composed of inert surface-crust material.*

*Above, collisions between planetoids could be the progenitors of meteoroid swarms of asteroidal origin.*

of the Geminids, for example, appear at lower altitudes (80 to 100 kilometers) than those of the Perseids (90 to 120 kilometers).

Note further that the Perseids' ZHR refers to their normal peak, but a secondary peak has appeared recently with an average ZHR of 220 over the past five years (see Chapter 6). The peak dates for the Perseids and the Leonids have moved ahead a few days compared with those cited for the 19th century due to the precession of the equinoxes. This is a motion that the Earth's axis traces over 25,000 years which causes, among other things, the stars' positions in the sky to change gradually. Naturally, the peak dates of other showers have shifted as well. If we refer back to medieval and ancient peaks, we see that the shift is even larger. For example, the Perseids' peak in 36 B.C., during the first recorded occurrence of this shower, was on July 15. The Bielids are not listed in the table because their activity is so slight as to be practically indistinguishable from that of sporadic meteors.

## Meteors From Asteroids?

A further look at the table on page 88 reveals the curious fact that the Geminids appear to be bound to the asteroid Phaethon, a so-called Earth-grazing asteroid—an object whose orbit intersects the Earth's and that could potentially become a lethal projectile in a collision with our planet. Thus not all meteors are the children of comets. The relationship between Phaethon and the Geminids was suggested by Fred Whipple in 1983, just after the asteroid was discovered, when he compared its orbit with that of the meteor stream. Bo Gustafson of the Max Planck Institute in Bonn confirmed the kinship in 1990 by analyzing in detail the orbits of Phaethon and 20 Geminids identified in the 1950s by Whipple and Luigi Jacchia of the Massachusetts Institute of Technology in Boston.

Also in 1990, Jack Drummond of the University of Arizona and Ian Halliday of the Astrophysics Institute in Ottawa compared the orbits of 139 Earth-grazing asteroids with 89 fireballs and found traces of another four streams of dust particles that seemed to originate with asteroids. We have yet to establish, however, which meteor showers may be associated with

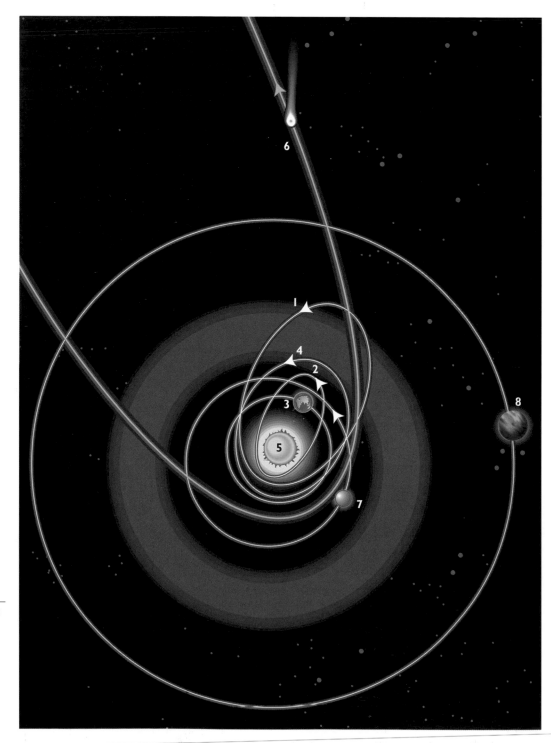

*The orbits of many asteroids that intersect the Earth's orbit resemble comet orbits. The gray belt is the region in which the great majority of asteroids orbit.*

*1. Adonis*
*2. Icarus*
*3. Earth*
*4. Apollo*
*5. Sun*
*6. Comet*
*7. Mars*
*8. Jupiter*

them. According to the two researchers, their peaks should fall on February 12, August 14, September 29 and December 22. Two of these dates coincide with the peaks of three known minor showers—the Beta Triangulids and Alpha Ursamajorids (August 14) and the Ursids (December 22).

It is still difficult to determine the mechanism that generates meteoroid particles from a body as compact as an asteroid. The dust could be the detritus from an asteroid's ultimate breakup or fragments produced by a collision among larger meteoroids released from the surface of an asteroid following a collision with smaller asteroids. Another hypothesis would link the debris to the heating that Phaethon suffers when it approaches perihelion—only 0.14 AU from the Sun, closer than any other asteroid.

In any case, Phaethon appears to have released the meteoroids over a period of several centuries, not all at once—much the way comets operate. And calculations demonstrate that if the particle release happened through a process of off-gassing similar to what happens at the surface of comets, the kinetic parameters would be completely compatible with the speed and direction of the Geminids observed. So does that make Phaethon a comet? No cometlike activity has in fact been observed, but it could be an extinct comet or one in a quiescent phase.

Ever since Whipple presented his model of a comet's nucleus in the 1950s, it has been supposed that with the passing of time, comets could cover themselves with an inactive crust. The images of Halley's nucleus taken from the Giotto probe have supported this hypothesis by showing how some regions of its surface

were covered by a black, inactive crust. And some short-period comets observed from Earth, like Neujmin 1 and Arend-Rigaux, seem to have become completely dark and inert.

Moreover, it is difficult to explain the existence of so many known Earth-grazing asteroids (several thousand at least) if they originated in the main asteroid belt between Mars and Jupiter. Their highly eccentric orbits, in effect, look more like those of such short-period comets as Encke. It seems, then, that many Earth-grazing asteroids now located within the orbit of Mars could eventually attain trajectories similar to those of asteroids in the main belt.

Add to this the presence of the Centaurs (see Chapter 2), and you begin to realize that comets and asteroids are more closely related

than has been supposed. The next few years may tell us what the boundaries between the two categories of object are.

Returning for a moment to the Geminids, we note yet another peculiarity—the fact that they do not have a tendency to break up during flight. This is readily explained if we remember that since they originated from an asteroid, they must possess greater density (about two grams per cubic centimeter) and compactness. Such asteroids would be comets in disguise. In addition to Phaethon, the asteroids Hidalgo and Oljato are two very likely candidates. On the other hand, it has also been observed that comets can evolve a trajectory similar to those of asteroids in the main belt from an orbit which was initially very eccentric, with a perihelion within the orbit of Mars.

*Trail of a particularly bright Geminid taken by amateur astronomer Rick Schmidt of Kansas.*

# A Date With Falling Stars

## Observing Meteors

In this chapter, we will prepare the reader for observing the principal meteor showers that occur over the course of a year. We can also, with a good dose of uncertainty, attempt a few predictions on the possibility of witnessing a really great show before the end of the millennium.

The only tool needed to observe meteors is the naked eye. A telescope doesn't serve the purpose because it restricts the field of view too much. The same can be said of binoculars; their gain in magnification simply does not compensate for what they lose in terms of visible field, although some stargazers do specialize in observing binocular meteors, which are too faint to follow with the naked eye.

Naturally, your observing must be done in as dark a place as possible, far from the lights of a city or town. You cannot hope to see meteors when you're standing under a streetlight or looking through your skylight at home! The ideal would be to get as far away as possible from any light pollution and climb a mountain at least 1,000 meters high.

To evaluate the sky's transparency, you can estimate the so-called limiting magnitude—the luminosity of the faintest stars visible—by using some of the stars in the constellation Ursa Minor for comparison. Ursa Minor is visible in every season of the year at a convenient height above the horizon. In conditions of perfect transparency (such as high in the mountains), the human eye can see stars to magnitude 6.5. Recalling that brightness increases 2.5 times from one magnitude to the next, you can immediately calculate that in a sky whose limiting magnitude is 5.5 (a decent country sky), you will see only 40 percent of the meteors that would be visible in a perfect sky. If the limiting magnitude is 4.5 (as in city suburbs), you will see only 16 percent, and so on.

Clothing is important too. Not every shower happens in the summer. For winter observing, it is important to dress to the max with, for ex-

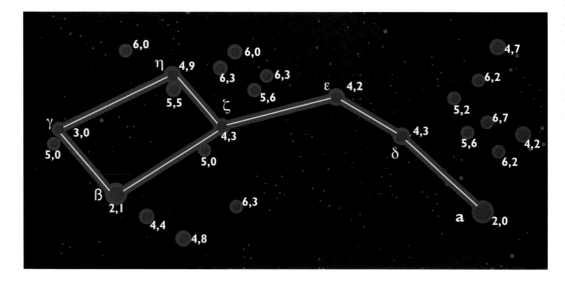

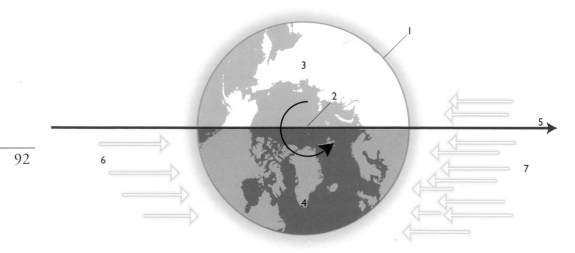

*Top, map of Ursa Minor shows comparative star magnitudes to one decimal place. To locate Ursa Minor, use the chart on page 105.*
*a. Polaris*

*Bottom, daily variations of meteor visibility depend on a combination of the Earth's orbital motion and its axial rotation.*
*1. Earth*
*2. North Pole*
*3. Day side*

ample, a down-filled parka, two or three layers of sweaters, leggings, lined pants, double socks, moon boots, gloves and a hat. You can drink hot beverages during an observing session, but absolutely nothing alcoholic. You might see "shooting stars," but after a momentary

happens after midnight. With regard to the periodic showers, keep in mind the effects of the combined rotational and revolutionary movements of the Earth. As it happens, meteoroids hitting Earth in the evening hours strike the part of our planet that is moving away from them. In the morning hours, on the other hand, the

4. Night side
5. Direction of Earth's orbital motion
6. Incoming meteoroids in the evening
7. Incoming meteoroids in the morning

Right, a fireball captured on the night of August 11/12, 1993, by Alberto Zinelli, Marco Amoretti, Luigi Mauilla and Leonardo Zanichelli of Collecchio, Italy, using a 28mm lens.

Earth is heading into the particle stream, in terms of both rotation and revolution. The same effect can be seen when insects splatter on a car's windshield but not on the back window or, in the case of rain, when water cascades violently over the windshield but leaves the back window almost dry. Obviously, sporadic meteors are not only less frequent but fainter too, because they fall at a lower geocentric velocity.

When you observe a meteor shower, don't be surprised if the meteors do not appear right near the radiant. This is the point from which perspective makes the trails appear to diverge —but only if they were extended backward. As noted earlier, meteoroids become incandescent only at a certain altitude, so the first part of their path through the atmosphere is often much longer than what is visible. Unless all the trails are accurately sketched on a star chart, it is sometimes difficult to see whether a given meteor belongs to the shower in question. This task is often taken on by observers belonging to organizations engaged in the close surveillance of meteors, such as the Meteor Division of the Unione Astrofili Italiani.

## Photographing Meteors

Photographing meteors is both easy and difficult. It is easy because the necessary apparatus is very simple: a camera with a "B" setting, a tripod, a flexible cable that locks and a sufficiently sensitive film, around 1000 ISO. The "B" setting holds the shutter open for as long as you like, and the cable fastens to the shutter release to control the exposure for the time needed with-

*Top, a fireball of magnitude −4 on the night of August 12/13, 1993. Photograph by Francesca Scarra, UAI Meteor Division.*

*Bottom, a fireball of magnitude −10 photographed in Prvic,*

*Croatia, on August 13, 1994. The wider-than-usual trail is due both to the object's great luminosity and to the use of a greater focal length (85mm lens in 6x6 format; equivalent to a 43mm in 24x36 format). Photograph by D. Sirovica.*

out trying to keep an unmoving finger on the button for the entire duration. But it is also difficult for two reasons: a meteor's light is not concentrated in one point, like that of the stars, but instead trails across a considerable area; and the phenomenon is almost too brief to be captured on film. The unfortunate result is that unlike most celestial phenomena, photographs of meteors are less spectacular than what you see by direct visual observation. Their trails look thinner, and their brightness is far less than what the eye perceives. This is all well illustrated by the images reproduced here, and it proves dramatically true of fireballs. These events are anything but rare. It has been calculated that you can see one fireball for every 200 hours of observing, but the frequency is 100 times greater during a meteor shower. Fireballs are uncommonly beautiful and, in the case of truly brilliant apparitions, altogether stupefying. What you see in photographs, unfortunately, is nothing more than an enlarged meteoric trail; this is well illustrated by the images in this chapter. Moreover, even meteors that seemed very bright at the time often leave no trace on film.

Remember that a camera's relative aperture, or f-number, is the ratio between the focal length of the lens (distance from the lens to the focus, the point where light rays converge) and the lens diameter. The smaller this number, the wider the lens is open and the greater the possibility of registering faint objects—like meteors—with a shorter exposure. Even so, we can hope to register only the brightest meteors— first magnitude or brighter—with this appara-

tus. It will be very difficult to reproduce their color too, because their trails do not leave enough of an imprint for a long enough time on a given area of the film to activate the layers that are sensitive to color. Moreover, we cannot delude ourselves that it will be easy to capture persistent trails, even the brightest ones or those which last the longest.

While this is not the place to detail them, there are various technical factors that make the camera's "normal" lens preferable. This is the 50mm lens, which usually has a relative aperture of f/1.8 (lens aperture of 28mm). Even better for our purposes, the normal lens functions at f/1.4 or f/1.2. Wide-angle lenses, such as 35mm, 28mm, 24mm and 20mm, are less ef-

ficient, both because of the smaller focal length (the capacity to register dim stars is directly proportional to the focal length) and because they do not open as far (usually to f/2.8). These disadvantages are distinct negatives in comparison with their only advantage—a larger field of view that allows you to keep a bigger piece of the sky under observation. An exception can be made for a fish-eye lens whose focus is 16mm or less, which usually also opens to f/2.8. Even if these lenses prove less efficient, they do permit the brightest meteors in the sky to register, and you are almost certain to capture any fireballs that might appear. In the case of telephoto lenses, on the other hand, the advantage of being able to capture fainter

Japanese amateur astronomer Yasuo Taguchi documented the remarkable activity of the 1991 Perseids. With a 34mm fish-eye lens in 6x7 format (equivalent to a 16mm in 24x36), he captured a good 26 meteors on the same frame. This exposure was held for at least 68 minutes by using less sensitive 400 ISO film and a lens at only f/4.5.

would be senseless to risk ruining such a rare shot by maybe prolonging the exposure and then finding the frame completely blank!

When photographing during a meteor shower, it's best not to point the camera directly at the radiant, because no meteor trails are normally found there. A good rule is to point the camera about 40 degrees above the horizon and 20 to 30 degrees to the right or left of the radiant. Good shots have been taken, however, with the camera pointed anywhere in the sky, even right at the radiant.

During a shower, it probably is not wasteful to use an entire roll of 36 exposures; who knows when another chance will come around? Neophytes should not be scared away if there are a lot of bright streaks on the exposure; neither should they think these are meteors. They are simply star trails—the streaks that stars leave due to the Earth's rotation on its axis. The longer the exposure, the longer the star trails. Meteor trails, on the other hand, will be more or less perpendicular to the star trails and far fewer in number (at least in the case of an ordinary shower; in a storm, the meteor trails would be as numerous as those of the stars). Naturally, if the camera is connected to a clock drive, it compensates for the Earth's rotation, so the stars appear as points of light.

Here again, as with the wide-field comet photography, you can embellish the shot with landscape elements such as mountains or trees, especially when photographing a shower whose radiant is low on the horizon.

### Dates You Cannot Miss

Now we want to give you some pointers on the nights when looking for meteors is especially

meteors is amply outweighed by the much smaller field of view. Nonetheless, this should not keep you from using much longer focal lengths to capture an occasional meteor in the course of photographing other objects.

Exposure times will be highly dependent on the sensitivity of the film, the type of camera lens and the state of the sky. Under a perfectly dark sky far from light sources and with no Moon, for example, your maximum exposure will be about 1 minute with a normal 50mm lens open to f/1.8 (like most commercial lenses) and using a 3200 ISO film like Scotchchrome 800-3200 developed for this sensitivity. Remember, too, that the film will tend to wash out as it registers the bright backdrop of the sky. Using a slower film around 1600 ISO and leav-

ing the other elements unchanged, exposure times can be increased to 3 to 5 minutes. With even slower films, around 400 ISO, you can push the duration of the exposure to 15 to 20 minutes. All the emulsions cited are color slide films. It is possible to make prints, of course, but slides are far better; they can be projected at a conference or for an evening among friends and, eventually, also printed. Naturally, the slower the film, the lower the chance of registering an event. Moreover, if sky conditions are not perfect, the exposures will have to be shorter (let's say, cut in half), because the film will tend to wash out too easily.

If a bright fireball happens to cross the camera's field, it would be a good idea to narrow the aperture and switch to a different exposure. It

*Fireball of magnitude −5 photographed with an all-sky camera on the night of August 11/12, 1993, by Marina Bolis. (UAI Meteor Division)*

worth the trouble because you will be sure to see at least a few dozen. We will also hazard a few predictions about the possible stormy return of the Leonids.

We will start with showers that are less dramatic and more predictable. As you can see in the table on page 88, these are the Quadrantids,

*Sometimes meteors casually appear in the field of view of the largest instruments. At top right, a brilliant meteor streaks past the Andromeda Galaxy (Schmidt telescope at Palomar). Top left, a fireball seems to pass through the nucleus of the galaxy NGC253, then increases tremendously in brightness before leaving the instrument's field of view (UK Schmidt at Siding Spring, Australia). The two instruments have the same diameter, 1.22 meters, and almost the same focal length, 3 meters. Bottom, fireball captured by Eraldo Guidolin on August 13, 1986.*

the Geminids and the Perseids. Their visibility is more or less the same from year to year. The optimal conditions are when the Moon is nearly new and cannot interfere with your observing; on average, this happens every three years. Unfortunately, the exact hour of the peak cannot be predicted with great certainty, but it is noted each year in publications like the *Observer's Handbook* of The Royal Astronomical Society of Canada. Aside from the Quadrantids and the secondary peak of the Perseids, peak activity usually lasts for several hours, if not days.

### The Quadrantids From the Lost Constellation

In the table on page 88, you can see that the Quadrantids have had a fairly high ZHR in the past few years. ZHRs of 170, 145 and 150 were recorded in 1992, 1993 and 1995, respectively. This means that under good conditions, you could see between 50 and 70 meteors an hour. Many brilliant meteors were observed in 1992 as well, including a few fireballs that were yellow-green in color and decisively brighter than Venus (magnitude –4.5). The shower's radiant lies between the head of the constellation Draco and the last star in Ursa Major's tail. Ursa Major is very easy to find. During that period, it lies very low in the north in the evening, while it is practically at the zenith before dawn. The radiant is circumpolar; it never sets in the course of the night, but it is very low and skims the northern horizon in early evening on January 3. Just before dawn, it sits high above the northwestern horizon, so this is the best time for observing. In effect, there is really no reason to leave the house before two in the morning.

Unfortunately, the Quadrantids' peak usually lasts for only a few hours. More specifically, once the peak is reached, activity drops by half in just four hours. And after all, the frigid January nights do not really elicit a desire to prolong one's observing session.

Incidentally, the orbit of the Quadrantid stream is changing rapidly. Its passes near Jupiter over the years have caused significant dispersion. While the stream intersected the Earth's orbit in the early centuries of our era, it then disappeared for more than a thousand years, until about 1700. In another three or four centuries, the Quadrantids will probably go missing again. These rapid orbital variations make it almost impossible to identify the stream's parent object, though there are many indications that it is a comet—the frequency of persistent trails, for example.

### The Geminids, Daughters of the Sun's Son

The Earth did not begin to intersect the orbit of the Geminid stream any earlier than the 19th century. This stream was only identified in 1862, and its activity increased throughout the 20th century. Recently, the Earth has probably been crossing the densest clump of particles released by the asteroid Phaethon, while around 2100, the stream will again be far from the Earth's orbit. The Geminids are even more predictable than the Quadrantids, although their observing conditions, such as climate, are quite similar (the Geminids' peak occurs around December 14). Due to the Geminid meteoroids' greater density and compactness and their relatively low geocentric speed, their trails last a little longer than those of other showers. The frequency of brilliant events is also high. All of this clearly favors the possibility of photographing the Geminids. The brightest ones often break into a chain of brilliant sparks that are very photogenic. Again, due to its stream's tighter composition, this shower produces several meteors whose trails persist in the sky.

The Geminids' ZHR has increased since the 1970s, when it was about 70. In the past five or six years, it has almost always gone over 100, and it will probably continue to increase for some time before declining. So we are currently guaranteed a chance to admire between 30 and 50 meteors an hour when skies are clear. Moreover, you need not wait until dawn, inasmuch as the radiant, which almost coincides with Castor in Gemini, is already high up at ten in the evening. To find Gemini, refer to the winter constellation of Orion, which is very easy to find because it is rich in bright stars. It lies due south and midway between the zenith and the horizon at midnight on December 14. You need only extend a line through the two brightest stars in Orion—Rigel and Betelgeuse—to find the constellation Gemini.

The best conditions for observing occur at 2 a.m., when the radiant moves to within a few degrees of the zenith. The stream is quite active from two days before the peak until one day after. In any case, the peak usually occurs on the night of December 13/14 and is well defined, followed by a rapid decline. The best dates for observing the Geminids, then, are essentially December 11, 12, 13 and 14. That is, of course, if there is no Moon.

### Fiery Tears

The other best-known meteor shower, the Perseids, has not generally been spectacular, and we have seen why. Yet in the summer of 1993, some astronomers threw caution to the wind and issued hyperbolic predictions.

*Facing page, meteor captured in a long-exposure photo (2 hours 30 minutes) by Alberto Zinelli on July 27, 1987.*

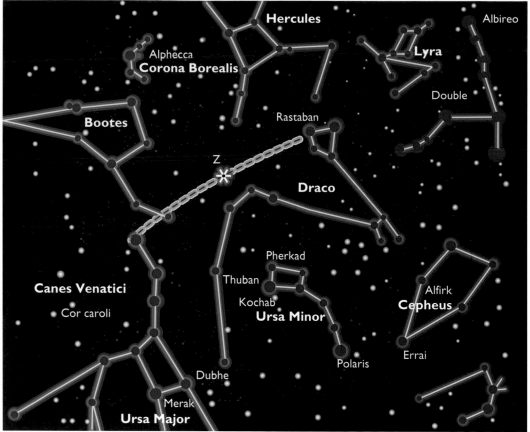

The truth is that the picture we have thus far sketched of the particle-stream distribution which gives rise to meteor showers—and, in particular, the Perseids—is too simplistic. In the years since 1988, a secondary peak has started to appear alongside the Perseids' primary peak activity. This happens a few hours earlier and recently reached a level greater than that of the primary peak. In 1991, Japanese sky-watchers observed more than 450 meteors an hour. In 1992, notwithstanding pronounced interference from a nearly full Moon and a very hazy sky, a peak of activity was seen in Asia such that, for a few brief seconds, a ZHR of nearly 8,000 was estimated. What's more, many

of the meteors were unusually bright; some even rivaled the light of the Moon. They frequently left persistent trails, with a tendency to fall in clusters, and many fragmented or exploded. Even a few Italian observers noted the stream's unusual activity just after dark. In Holland, several fireballs were seen at twilight, but by then, peak activity had abated.

All the conditions pointed to the fact that something new was happening. It seemed clear that this new peak derived from a dense stream of meteoroids recently released from the parent comet. This was proved both by the greater number of meteors and especially by the larger size of the debris, which gave rise to the bright

fireballs. Indeed, when such material detaches from the parent comet, there are often some rather large pieces. These fragment very slowly through reciprocal collisions, the effects of the solar wind and the thermal contraction and expansion caused by a continual variation in distance from the Sun. Now, since these just-released particles cannot have traveled very far from the comet's nucleus, it follows that the comet was not very far away and was probably about to return. This is, in fact, what happened.

Comet Swift-Tuttle arrived a little late. At the time of its discovery, around 1862, its period was thought to be 120 years. It was therefore expected to return in the 1980s—and a sudden

*Above left, a Geminid crosses the constellation Taurus during the shower of 1990. The planet Mars is visible as the brightest "star" below the Pleiades; and the Hyades star cluster, shaped like an upside-down V, can be seen at*

*right. Photograph by Terence Dickinson.*

*Above right, how to find the radiant (Z) for the Quadrantids.*

*Facing page, locator chart showing the radiant (Z) for the Geminids.*

growth in the Perseids' activity in August 1980, with a ZHR of about 200, did cause people to think that the comet had come back.

But Marsden suspected that the earlier calculations were mistaken and that the orbit had undergone some modifications, caused by an enormous rocket effect produced by violent jets of dust and gas that the most active comets emit when close to the Sun. This activity has the effect of either slowing down or accelerating the comet and lengthening or shortening its period.

Marsden thus proposed a period of about 130 years—varying from one visit to another due to the effect mentioned above—and November 25, 1992, as the date when it would reach perihelion. On September 26 of that year, in fact, Japanese comet hunter Kiuchi found the comet again as it was approaching perihelion on December 12, substantially confirming the predictions.

Did all this suffice to feed the notion that there would be a meteor storm in 1993? Certainly not, but it was enough to go back and review events in and around 1862, the date of Swift-Tuttle's previous visit. In 1861 and 1862, there were reports in the Far East of a remarkable increase in the Perseids' activity. There

were also numerous observations in the West on August 10, 1863. While leading a team of six observers in Connecticut, H.A. Newton saw 153 meteors in half an hour under conditions that were inferior due to interference from moonlight. In Germany, Eduard Heis, a renowned meteor researcher, wrote that at a certain point, the faintest meteors could no longer be counted. The bright ones alone reached 166 an hour. Another German observer reported a maximum hourly rate of 216. In Rome, astronomer Caterina Scarpellini reported 197 meteors in

the entire night but admitted that she, too, had not documented the faintest ones, which were impossible to count. Julius Schmidt, director of the Athens Observatory, was traveling on the Ionic Sea and estimated an hourly rate of 113. He also observed several brilliant fireballs.

In short, aside from some rather vague references to an ample number of falls, the Perseids reached a maximum hourly rate of somewhere between 200 (in Europe) and 300 (in the United States), or about three to four times the normal rate. Not bad, but certainly no storm.

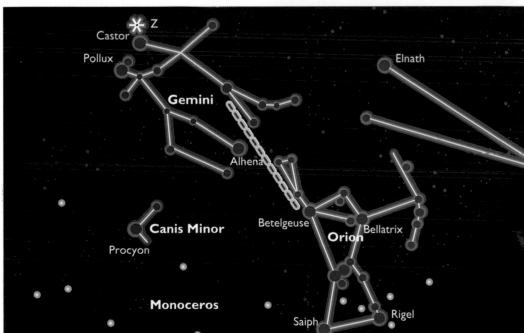

*Above left, an absolutely splendid fireball appeared in 1860; it is depicted here in an 1862 painting by F.E. Church.*

*Above right, Comet Swift-Tuttle on its 1992 visit, photographed on November 23 at the Cortina Observatory by Alessandro Dimai.*

Nevertheless, the situation seemed different in 1993. According to Joe Rao, a noted U.S. meteor expert, the stream of matter recently ejected from Swift-Tuttle's nucleus—which had begun to renew its rendezvous with the Tears of St. Lawrence in 1991 and 1992—had never before encountered the Earth. This is because the comet's orbit changes slowly over time, due to gravitational perturbations by the planets. Thus in 1993, Swift-Tuttle's orbit and the Earth's orbit reached a minimum separation of only 140,000 kilometers, compared with 750,000 kilometers in the 19th century and 3.5 million kilometers in the 18th century.

Around 1737 (the date of an earlier Swift-Tuttle apparition), the Earth was very far from the comet's orbit and therefore from the meteoroid stream that should have been released right then, and no extraordinary event was recorded. Around 1862, the Earth skimmed the stream, leading to the increased activity cited earlier. In 1991 and 1992, the Earth started to penetrate the stream, and in 1993, as it moved into the middle of the stream, something truly exceptional was about to happen.

There had been other favorable factors in 1862. When the Earth met the comet's orbit on that pass, it was closer to the comet's nucleus and therefore closer to a zone that was potentially rich in dust. Furthermore, Swift-Tuttle produces strong jets during every near approach to the Sun (as we can deduce from the intensity of the resulting "rocket effect").

Again, was all of this enough to predict a meteor storm? No. In fact, Rao and others were only making a hypothesis that something like the event observed in 1863 was more than possible, even probable. According to Rao, a peak would occur at 1:15 Greenwich time (3:15 in Italy) on the morning of August 12, an hour that favored European observers but would impede North American and Asian viewers. The whole thing was expected to last for one to two hours at most. Unfortunately, the Italian mass media overblew the forecasts and gave their readers an impression of near certainty that a considerable meteor shower would occur. This produced a veritable exodus from cities and towns into the surrounding countryside, hills and mountains, which seethed that night with thousands upon thousands of curiosity seekers and enthusiasts anticipating a great celestial spectacle.

Few people, however, waited until the time predicted for the peak. It was August 11, and members of the Feltre Astronomical Association *Rheticus* were on hand to share the wonders of the firmament with about 3,000 people

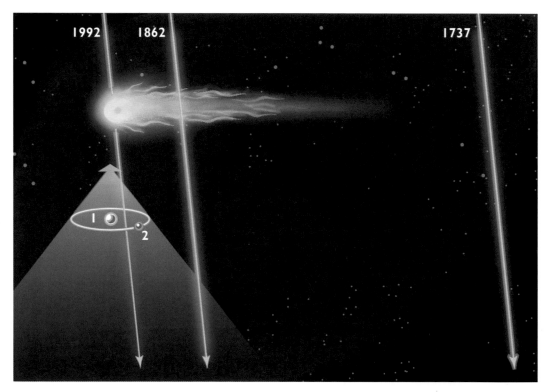

Top, relative separation between the orbits of the Earth and Comet Swift-Tuttle during the three historically recorded visits. The Moon's orbit is shown for comparison.
1. Earth
2. Moon

Bottom, this illustration of a spectacular fireball that appeared in 1863 was published in an 1863 bulletin of the Royal Belgian Academy of Science along with an account of the Perseids' activity that year.

gathered on the 1,454-meter summit of Mount Avena. Few of us were left after midnight, however, to witness the explosion of fireworks.

Before the evening began, the sky was completely covered by a mass of clouds, which happily dissipated around 9:45, just as darkness fell. The sky, nothing exceptional to begin with, became clearer and clearer, while the mist deposited itself on the ground. A number of shooting stars began to furrow the sky, some of them very bright—true and proper fireballs that elicited applause and fervent shouts of admiration. At 10:30, with the radiant in Perseus already rather high, the meteors multiplied and gave us good reason to hope. But in the following hour, even though the sky became ever clearer, it soon became evident that the shower's activity had begun to diminish rather than increase.

At midnight, the sky indecorously covered up again. Everyone waited patiently, a half-hour, an hour, an hour and a quarter. Suddenly, a shudder raced through the encampment. A livid flash blazed through the clouds, leaving us dumbstruck. It had to be a fireball as bright as the full Moon! People elsewhere in Italy who observed it under a clear sky confirmed our assessment the next day.

At 1:30 a.m., fatigue and sleepiness got the upper hand, and almost everyone left. Too bad. At 2:00, the mist again drifted to the ground and left an even clearer sky. This was important, because the Moon had now been up for an hour and a half, and its last-quarter light was a serious constraint. But the clarity of the sky continued to grow, and the Moon caused little disturbance outside its own small zone. In any case, the radiant was far enough away from our satellite by then and was riding very high.

At 2:30, in an almost fairy-tale landscape, with a field that earlier had been full of telescopes now nearly deserted, about a dozen remaining people were contemplating the celestial canopy, overwhelmed. Some were standing, others seated at their telescopes, still others stretched out on blankets or sleeping bags. Then something moved. We knew it. We believed it. Rao got it right, right down to the hour.

The trails multiplied, they got brighter, and we began to see a lot of fireballs. In the preceding hours, we saw only about one meteor a minute; now the count rose to two, three, five. At 3:05, a group of three and then five observers began to count, but only for short stretches. We were too tired to face an extended count. We limited ourselves to five-minute periods, and during one of those counts, we recorded 25 meteors in

*The orbits of Comet Swift-Tuttle (1) and Earth (2) around the Sun (4) are almost perpendicular. The blue lines represent the orbits of particles of the Perseid stream (3) that produce the primary peak. The tiny white ellipse at the intersection of the orbits corresponds to the area covered by the diagram on page 102.*

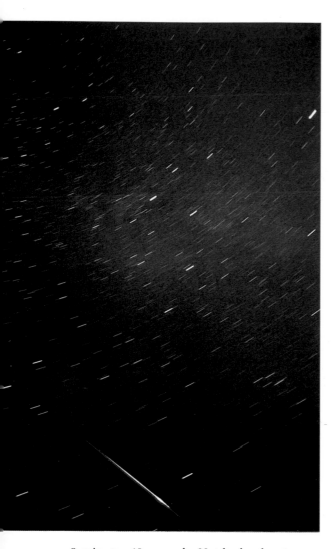

3 minutes 40 seconds. Not bad—about seven per minute. We suspended the count at dawn, which broke at 4:25, when the intensity of the shower seemed to be dropping off. We counted 112 meteors, but in two hours, we saw at least 500. We were satisfied. It wasn't a storm, but none of us had ever seen anything like it.

According to the data coordinated worldwide by the International Meteor Organization (IMO), which gathered hundreds of observers' reports, the maximum ZHR occurred at around

3:30 a.m. Greenwich time (5:30 in Italy, when it was already quite light), and it peaked at 300 meteors per hour. This means that the luckiest observers effectively saw about 150 meteors an hour under clear skies. In the end, it seems that everything happened much as it had during the 1870s. The maximum ZHR we recorded at about 1:45 a.m. Greenwich time was much higher, at approximately 750; but this was spurious. Our counts were not lengthy, and we had the impression that more meteors fell when we were not counting. Moreover, to calculate the ZHR properly, the count has to be taken by a single observer, and we were counting as a group. In terms of the first point, however, friends from an astrophiles' association in Vicenza confirmed our results. They observed continuously for the entire evening. From 10:00 p.m. on August 11 to 2:45 a.m. on August 12, their hourly rate held constant at 60 to 70 meteors. From 2:45 to 4:15, 321 meteors were observed, giving an hourly rate of 214. After 4:15, they also counted fewer meteors (39 in a quarter-hour), as if the shower itself was now exhausted. Considering that their sky was less clear than ours and that they were observing on flat ground, their ZHR would really double to nearly 1,400 and would thus confirm our impression of a higher number during the times when we were not counting.

In terms of the second point, it is difficult to judge how counts taken by one person and by more than one can be compared. Neil Bone, director of the Meteor Division of the British Astronomical Association, however, proposes some reasonable corrective factors. For five observers, it would be necessary to multiply the data by 0.32; for three observers, by 0.41. Taking an average, our ZHR would be about 270, in

line with the IMO data. The more trustworthy count performed by the people from Vicenza remains decidedly higher, about 500. And, in fact, the data gathered nationally by the Meteor Division of the Unione Astrofili Italiani place the ZHR between 500 and 600. In successive years, the activity has continued higher than normal, though less than outstanding. The ZHR was 250 in 1994, 160 in 1995 and 155 in 1996. It now appears that the secondary peak has already been exhausted. Nevertheless, the Perseids' primary peak, which usually falls on the night of August 12/13, is normally very regular, like that of the Quadrantids and the Geminids. Because the weather is so favorable then, among other things, the Perseids are the only commonly known shower—so much so that some people think meteors can be seen only in August. As we have described, however, they can be seen in other seasons, too, and with the same intensity, especially in the winter months. But who spends their evenings outdoors in December or January?

The Perseids' radiant is very close to the star Eta Persei. The constellation Perseus is located next to Cassiopeia, which can be found by starting from the Big Dipper, the well-known configuration that is part of Ursa Major (the Great Bear). To find Polaris (the North Star), draw an imaginary line between the last two

*Above, a meteor of magnitude −3 photographed on the night of August 11/12, 1993, by Maurizio Eltri. (UAI Meteor Division)*

*Right, three meteors captured in the same frame on August 11/12, 1993, by Alberto Zinelli, Marco Amoretti and Leonardo Zanichelli.*

stars of the Dipper, then extend the line five times that distance. Now draw a line toward Polaris from the first star of the Dipper's handle and extend it the same amount past Polaris on the other side to find Cassiopeia.

Around 10:30 p.m. on the date of the peak, or as soon as it's dark on the night of August 12/13, Ursa Major lies low in the northwest and Cassiopeia is low in the northeast. The radiant is very low, and it rises slowly in the course of the night. Just before morning twilight begins, around 4:30 a.m., the radiant reaches its peak altitude at roughly 10 degrees from the zenith. The shower activity typically lasts for about three days, from August 10 to 13, during which a ZHR of 40 to 50 is produced. It doubles with the arrival of the peak, which lasts for a few hours. The decline of the peak is much faster than its rise. A high percentage of the Perseids are very bright and leave persistent trails, the trait that makes them such attractive photographic subjects.

**The Draconids, or the Possible Surprise**

As shown by their omission from the table on page 88, the Draconids (also called the Giacobinids, from the name of one of the discoverers of their parent comet) do not belong among the major showers. All the same, this shower occasionally gives rise to remarkable episodes that coincide with the parent comet's approach to perihelion. This is Comet Giacobini-Zinner, which has a period of 6.4 years. On October 9, 1933, there was an absolutely unforeseen shower visible over most of central Europe, and therefore in Italy as well. We spoke with an eyewitness to the event, who found the sight absolutely spellbinding. He thought the frequency of falls reached 900 *per minute* for a

short time! According to some, it was the biggest meteor storm of the 20th century.

There was another great shower on October 9, 1946. Notwithstanding the nearly full Moon, an hourly rate of at least 6,000 meteors was recorded. Those detected by radio signals were still more frequent, reaching a peak of 170 a

minute. Despite the limited sensitivity of the photographic films of the day, Carl Seyfert succeeded in capturing at least 40 meteors in one 12-minute exposure.

Generally, however, the Draconids' ZHR falls under 10. Only two significant episodes of anomalous activity have been recorded. In

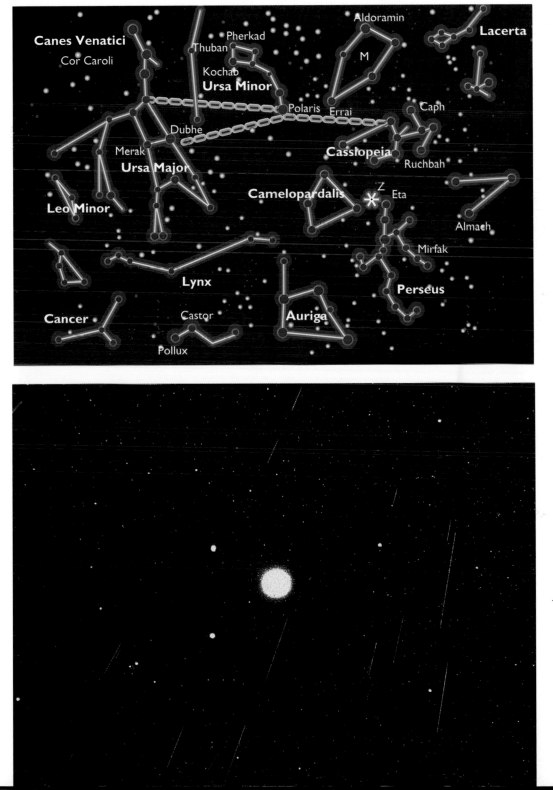

*Top, locator chart showing the radiant (Z) for the Perseids.*

*Bottom, exceptional document of the Draconid storm of 1933. At least 20 meteor*

*trails are visible in this 52-minute exposure. The field is centered on the star Vega. Photograph by F. Quenisset at Flammarion Observatory, Juvisy. (From* Astronomie populaire *by Camille Flammarion, 1955 edition)*

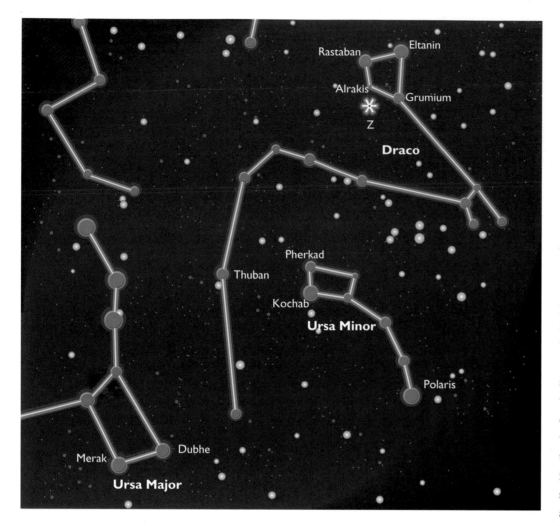

Labels on chart: Rastaban, Eltanin, Alrakis, Grumium, Z, Draco, Pherkad, Thuban, Kochab, Ursa Minor, Polaris, Merak, Dubhe, Ursa Major

ahead of or behind the comet, our planet may intercept a substantial quantity of comet dust. But not even this factor is conclusive in and of itself (as we will see later when we discuss the Leonids).

In 1933, for example, the Earth reached the node 80 days after Comet Giacobini-Zinner had passed by. During the 1933 shower, several meteors brighter than Jupiter and Venus were seen. In 1939, our planet flew by several months ahead of time with respect to the comet, while the 1946 passage occurred 15 days later. There were unfavorable circumstances again in 1959 and 1965, but the 1972 crossing occurred only 59 days after the comet, without anything in particular happening. The distances were high again in 1978, while in 1985, the Earth reached the node only 26 days after Giacobini-Zinner.

In 1992, our planet crossed the node at least six months after the comet. In 1998, however, the Earth crossed the stream 44 days before the comet's arrival. The lunar phase on the date of the peak, the night of October 8/9, was rather unfavorable. The full Moon fell on October 5, interfering with nearly the entire night of observing on October 8/9. However, an increase in altitude would have sufficed to make the moonlight a little less disturbing. Wherever the air is more transparent, in fact, atmospheric diffusion of the moonlight is less noticeable. At 2,000 meters, for example, the clarity of the sky away from the Moon in this phase is entirely acceptable.

The Draconids' radiant (next to the head of the Dragon) was at maximum altitude just after sunset on October 8 and was visible all night. What observers saw that night, as it turned out, was an average Draconid performance.

Notwithstanding their modest geocentric ve-

1952, the ZHR was about 200, while in 1985, it reached a rate of 730. In the latter case, a peak of 1,000 meteors per hour was reached during the day but was registered only by radio instruments.

It is not impossible that Giacobini-Zinner's most recent approach to perihelion in April 1998 could produce a noticeable increase in activity of the Draconids. But we should note that if the return of the parent comet were sufficient

to create great meteor showers, they would not be so rare. One example will suffice for all. When Halley's Comet returned in 1986, the Eta Aquarid and Orionid meteors associated with it showed no substantial increase in activity.

What does play a profound role is the moment when the Earth reaches the node—the point where it meets the orbit of the comet, with respect to the position of the comet itself. If it arrives at this point within a few dozen days

*Locator chart showing how to find the radiant (Z) for the Draconids.*

*Facing page, top, geometry of the 1966 meeting between Earth and the swarm of particles which caused the famous Leonid storm that year.*
*1. Leonid swarm*
*2. Earth's orbit*

*Facing page, bottom, Leonid storm of 1966 photographed by A. Scott Murrell of New Mexico with a 50mm lens at f/1.9. At least 100 meteors are visible. The exposure time was 10 minutes.*

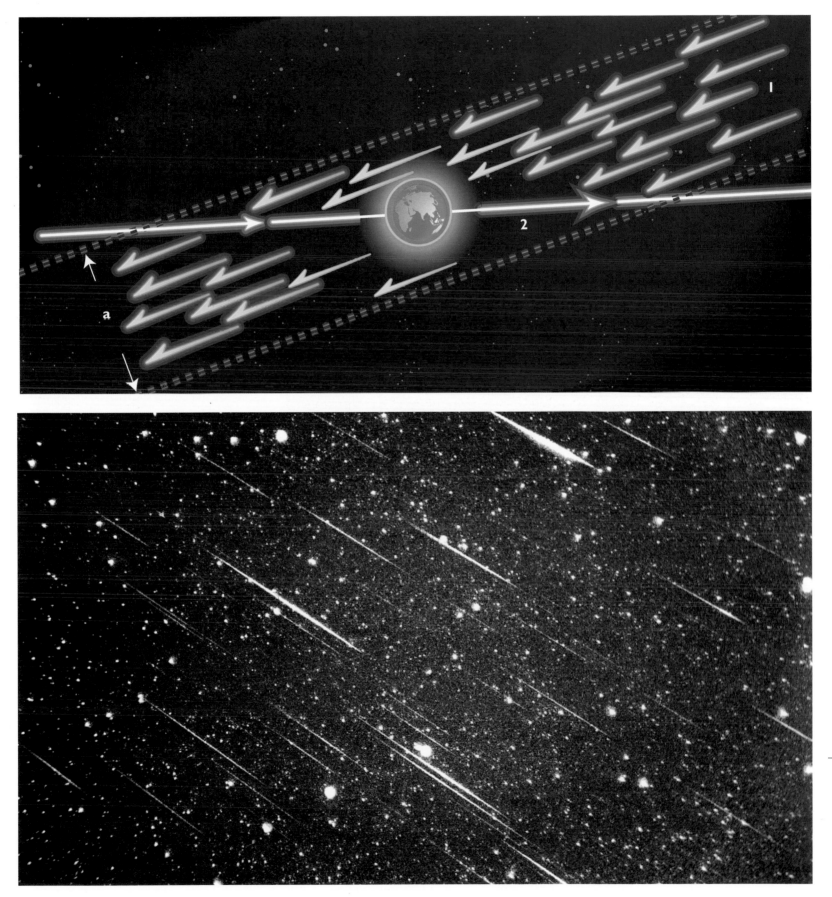

locity of 20 kilometers per second, the Draconids often tend to leave bright, persistent trails. According to Neil Bone, this may be explained by assuming that the meteoroids were recently released from the parent comet and may therefore still contain volatile frozen substances.

### Time for the Stormy Leonids

The Leonids produced only minor activity in the 1930s, similar to their performance at the turn of the 18th and 19th centuries. Their hourly rate achieved peaks of 190 in 1931 and 240 in 1932, but then nothing more. And no one detected Tempel-Tuttle as it approached perihelion again. You might imagine that it had met the same fate as Comet von Biela, dispersing its dust along a different trajectory, or that the orbit of the meteoroids had been modified by perturbations from the major planets so that the densest part of the stream would never again swing close to Earth.

As it turned out, the astronomers had made another mistake. The Leonids came back to life in the 1960s. There were 50 meteors per hour in 1961. In 1962 and 1963, there were 15 to 20, but there were 40 in 1964. In 1965, a rate of 120 per hour was recorded, with several fireballs brighter than Venus, whose trails lasted for several seconds. Moreover, that same year, Comet Tempel-Tuttle, which had been lost for almost a century, was recovered. The following year, however, no one expected much. But here's what happened on the night of November 16/17, 1966, as recounted by eyewitness Ken Croswell, a well-known American astronomy popularizer (once again, as in 1799 and 1833, the Americas were favored—this time the central and western United States):

"It is midnight, and Leo is rising. The beast's enormous head appears over the northeast horizon, followed a few minutes later by its bluish white heart, the brilliant Regulus. Every so often, a meteor sails through the sky over my head. Leo rises and Regulus shines brightly, but its activity is still weak. At 2:00, the rate is 30 an hour. The meteors are fast, and many of them are brilliant. Some are green or blue. I take note of everything—the color, the luminosity, the trajectory. They are the Leonids, but so few in number that I begin to think the experts were right.

"At 3:00, the constellation is high in the east, and finally, the peak builds—one, then two, then three a minute.

"A sense of excitement grips me. What do you

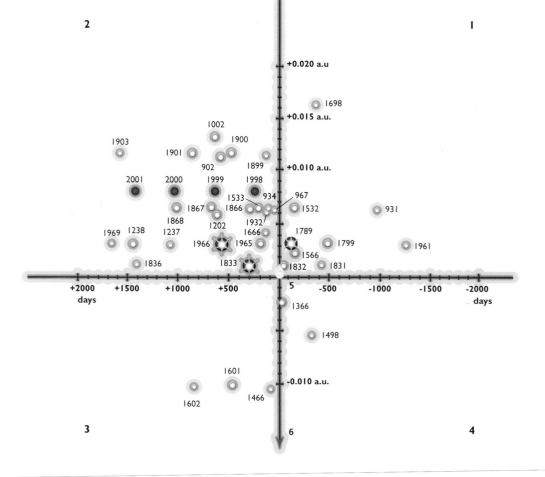

**MAJOR METEOR STORMS OF THE LAST 200 YEARS**

| Shower | Year | Hourly Rate* | Places Observed |
|---|---|---|---|
| Leonids | 1799 | 30,000 | South America, Florida, British Isles, Germany |
| Leonids | 1832 | 20,000 | Urals, Arabia, Mauritius, North Atlantic, Europe |
| Leonids | 1833 | 25,000 | Central and North America |
| Leonids | 1866 | 10,000 | Europe |
| Leonids | 1867 | 1,000 | North America |
| Leonids | 1868 | 1,000 | North America |
| Andromedids | 1872 | 6,000 | Europe |
| Andromedids | 1885 | 75,000 | Europe |
| Leonids | 1900 | 1,000 | Canada |
| Leonids | 1901 | 2,000 | United States, Mexico |
| Draconids | 1933 | 54,000 | Europe |
| Draconids | 1946 | 10,000 | Canada, England |
| Leonids | 1966 | 150,000 | United States |
| Draconids | 1985 | 1,000 | England, Japan |

*\* Refers to meteors actually seen or estimated, not to the theoretical ZHR. The figures for the 1946 and 1985 Draconids come from radio counts.*

The above table gives a compendium of the most spectacular meteor storms observed over the last two centuries.

Left, diagram of the possible geometry of meetings between the Earth and the comet dust released by Tempel-Tuttle. The vertical axis shows the swarm's distance in astronomical units (a.u.) outside (+) or inside (−)

want to bet that a great storm like the ones in 1833 and 1866 is about to arrive? Several big fireballs shoot across the sky, exploding silently in a white flash. The shower intensifies. There must be a dozen or more stars falling every minute, and I struggle to count them.

"Around 4:30, the shower becomes a furious storm, then a whirling deluge. Hundreds, then thousands of meteors shoot across the sky. I can barely estimate their number. Five a second? Ten? Twenty? It's impossible! 72,000 an hour?!

"At 5:00, Leo is high above my head and is vomiting meteor upon meteor. They rip through the sky at high speed, and I have to plant myself firmly on the ground to watch them. As when you drive a car through a nocturnal snowfall, I feel as though I am hurtling through space at maximum speed, flattened by the dense cloud of particles that the Earth is crossing through.

"The storm lasts an hour, then diminishes in intensity. As dawn arrives and clouds in the east grow red and purple, the shower is still producing several meteors a minute. When the sun rises, I am exhausted but happy. Whoever, like me, ignored the experts and kept a faithful vigil that night witnessed the biggest meteor shower of the 20th century."

Estimates of the hourly rate were highly variable, from 10 to 200 *per second* (720,000 an hour!). Reasonably credible estimates speak of 150,000 an hour, although in 1995, NASA's Peter Jenniskens criticized this number as excessive. He maintains that a rate of about 15,000 per hour is far more realistic based on radio observations made at the Springhill Meteor Observatory near Ottawa.

the comet's orbit as it passes us on its orbital plane. The horizontal axis shows the interval in days between the dust swarm and the parent comet as the swarm precedes (–) or follows (+) the comet. The larger yellow dots indi-

cate the years when there was a meteor storm (the most abundant are largest), while the smaller yellow dots indicate a year of minor but still remarkable activity. The blue circles show where the Earth crosses the orbital

plane of the comet between 1998 and 2001.
1. Meteoroids ahead of the comet and outside its orbit
2. Meteoroids behind the comet and outside its orbit
3. Meteoroids behind the comet and inside its orbit

4. Meteoroids ahead of the comet and inside its orbit
5. Comet Tempel-Tuttle
6. Toward the Sun

Above, photo of 1966 Leonid storm taken by Dennis Milan from Kitt Peak, Arizona, with a 105mm lens at f/3.5 in 6x6 format (equivalent to a 50mm in 24x36 format). No fewer than 70 meteor trails can be seen in this 3.5-

minute exposure. Notice in the center of the photo the pointlike trails left by two meteors that were approaching the observer head-on.

Jenniskens' work provoked a nettled response from Bradford Smith of the Hilo Institute of Astronomy in Hawaii, who supported the more optimistic count. In a letter to *Sky & Telescope* magazine, Smith wrote, among other things: "The authors of the radio measurements say that the antenna was saturated during peak activity and the observations had to be corrected by superimposed echoes. Therefore, there is no indication that the radio observations correlate with the visual ones. . . . My estimate of the hourly peak was, obviously, indirect. With the sky full of meteors and persistent trails, counting the total number visible at every moment was inconceivable. Instead, I estimated the number of meteors originating every second within a selected area of a known size, then followed it with the appropriate geometric correction. . . . While I can entertain being mistaken by some factor, I am certain that the frequency of meteors cannot have been as low as Jenniskens suggested—four meteors per second. Every time I recall that memorable morning, I believe that we were dealing with a number at least 10 times higher. There had to be."

In 1969, the Leonids produced another unexpected burst of activity; about 240 an hour were recorded in the northeastern United States. The advent of the 1966 storm after two blanks had been fired naturally heated up expectations for the parent Comet Tempel-Tuttle's approach to perihelion in 1998. Would we see anything? And who would be the lucky ones? The Americans again? The Japanese? The Europeans? It would be hard, if not impossible, to say.

First of all, the Leonids' activity typically increases during the six or seven years before and after Tempel-Tuttle approaches perihelion. The ZHR continued to grow in the years before 1998, as if announcing something truly significant. After 1969, the Leonids stabilized at a ZHR of around 15, up until 1991—an increase to 35 was recorded that year. In 1992, the Moon was at last quarter and not very bright, but it was close to the radiant (about 15 degrees), so it interfered with observing. In 1993, another ZHR of 35 was recorded. In 1994, it increased to 70 and in 1995 to 80.

In 1996, the Leonids' ZHR reached nearly 100, rivaling the activity of the Perseids, Geminids and Quadrantids. That year, there were also a lot of brilliant fireballs with lasting trails. One was even as bright as the full Moon and left a trail that lasted for six minutes. Another left a luminescent trace in the sky for 10 minutes. All that seemed auspicious, given that it followed in the footsteps of what happened in 1961, five years before the storm of 1966 (although something of the kind had happened in 1930 without this grand prelude but with a worthy conclusion).

In 1981, Donald Yeomans accurately analyzed the historic records of the Leonid shower and used them to map the distribution of dust around the comet, thus identifying the conditions most favorable to the production of a big shower. As we have noted previously, the shower has been observed since 902; yet over the past 1,000 years, we have not always had abundant showers. For one to happen, according to Yeomans, the Earth must pass through the part of the stream that lies on the outside of the comet's orbit and behind the comet's position, thus encountering the dust that is blown away by solar-radiation pressure.

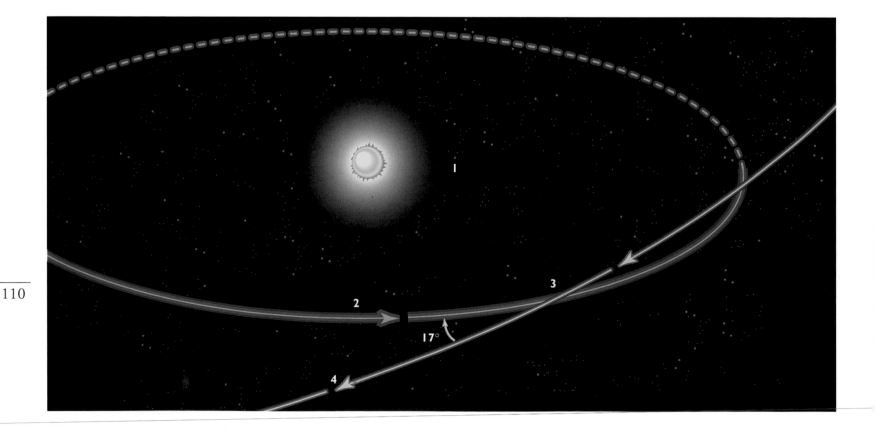

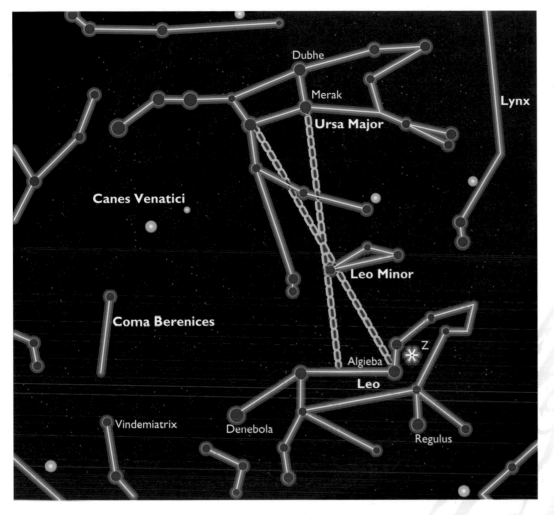

one. The one it met in 1966, for example, was 35,000 kilometers wide, and the Earth crossed it in one hour. Consequently, if the shower is visible in one location, it will be invisible in another. This situation is more critical for the Leonids than for other showers. In fact, the radiant is not adequately visible before 1 a.m. on the date of the peak, November 17. So there are only 4.5 hours before dawn in which we can hope that a shower will happen.

Unfortunately, the Earth is currently traveling farther away from the stream than it was at the time of the great showers of the last two centuries. This distance was about 480,000 kilometers in 1799; 180,000 kilometers in 1833; 970,000 kilometers in 1866; and 460,000 kilometers in 1966. By 1998-99, the distance had grown to 1,120,000 kilometers. On the other hand, the distance in 1932 was similar to that of 1866, and nothing especially noteworthy happened, so we would suggest that even this circumstance by itself is not decisive.

Let's posit that the swarm of dust particles giving rise to the hoped-for storm near the end of the millennium lies on the plane of the comet and note that Tempel-Tuttle, which was recovered in March 1997, reached perihelion in February 1998. Yeomans risked a few predictions for the time periods in 1998 and 1999 when a great shower could happen. These moments coincide with the Earth crossing the plane of the dust stream (the orbital plane of the comet).

In 1998, according to Yeomans, the Earth would pass the plane of the comet's orbit on November 17 around 8:45 p.m. Italian time, exactly 256 days after Tempel-Tuttle had gone by. But even if there were a large swarm of meteoric material at that point, nothing would be seen in Europe, since the radiant would not rise for another two hours

In fact, when the Earth met the dust lying inside the comet's orbit between 1266 and 1499, there was only one spectacular shower, in 1366. Since 1666, however, when the geometry of the meeting was what Yeomans describes, there have been numerous big showers. This favorable alignment occurred again in 1998–99, just as it happened in 1833, 1866 and 1966.

But this is not enough, either. The geometry was the same in 1900–01 and 1931–32, but activity still fell below expectations. In 1799, howev-

er, the Earth crossed the stream's orbit ahead of the comet, and a spectacular shower took place.

This irregular behavior, according to Yeomans, is easy to explain. The dust is distributed around the comet in irregular clumps, or swarms, of varying density, augmented by the inevitable release of additional dust from the surface of the comet's nucleus. A significant shower can happen only if our planet crosses one of the denser swarms. Moreover, these swarms are small, so the Earth takes an hour or two at most to cross

Facing page, orbits of Earth
and Comet Tempel-Tuttle.
The comet's orbit is tilted 17
degrees with respect to the
Earth's.
1. Sun
2. Earth's orbit
3. Position of Earth around

November 17/18
4. Tempel-Tuttle's orbit

Above, locator chart showing
how to find the radiant (Z)
for the Leonids.

(unless the storm were unusually long, as it was in 1833). The Asiatic continent would be favored according to this prediction—and justly so, since none of the great Leonid showers of the last two centuries was visible with Leo above the horizon during the night hours in Asia. Moreover, the Moon's phase would be extremely favorable (nearly new).

As the night of November 17, 1998, began to darken, astronomers around the world were eagerly awaiting the show. Many scientists, heeding the predictions, had packed up their gear and headed to locations in the Far East, hoping for the best view. Ultimately, what transpired was a truly outstanding meteor shower, but rather short of a storm by most accounts. Observers in western Europe recorded 100 to 300 meteors per hour in the predawn hours; North Americans saw 50 to 100. All observers reported an exceptionally high number of bright fireballs.

And what does the picture look like for 1999? The Earth is expected to reach the node at 2:50 a.m. on November 18, 1999—622 days after the comet passed by. In comparison, the Earth passed the node 561 days after the comet in 1966. If we then encounter comet dust, Leo will be ideally positioned in European skies, but past experience shows that it is difficult to predict where the most favored location will actually be. The Moon will be rather bright—a few days past first quarter—but it will set by midnight, leaving the sky completely dark before the predicted show. Your best bet is to head out to a dark site on the nights of November 16 and 17 (and possibly the 18th as well) and plan to stay awake into the predawn hours.

It is again possible, but not very probable, that a storm will happen in the year 2000, because the Earth will be too far from Tempel-Tuttle (passing the node at least 987 days earlier). In any case, the predicted time of arrival at the node on November 17, 2000, would likely favor the eastern United States, South America and the Atlantic. The Moon, however, will interfere considerably that year; it will be at last quarter, at about 20 degrees from the radiant, and will remain in the sky for most of the night.

Naturally, all this is mere hypothesis—reasonable, but not all that well founded. It is likely, instead, that the Leonid stream will be somewhat removed from the orbital plane of Tempel-Tuttle, and in this case, the peak could arrive several hours early or late. Among the more recent cases, the peak in 1965 was 13 hours ahead, the storm of 1966 was late by one hour and the shower of 1969 came four hours late.

Thus the schedules given are not to be taken as gospel. Rather, to avoid missing a unique event by a few hours, it is best to observe for the whole night, starting before the radiant rises, since the meteors that travel west will be visible even with Leo below the horizon, as happened in 1866.

There may be no storm at all. Nonetheless, it is worth watching on those nights. However this round goes, the Leonids will definitely, one of these years, produce a count of meteors superior to any other shower. If it goes well, everyone who witnesses that magical night will experience the inimitable feeling of literally being aboard a cosmic starship as it plows through space at a vertiginous speed, the combined movement of our planet and the particles producing a velocity more than double the 108,000 kilometers per hour at which the Earth flies around the Sun.

It would be madness to risk losing an opportunity to witness such an event—the first in 30 years in this part of the world. Indeed, according to the calculations of Marsden, Yeomans and Gareth Williams of the Minor Planet Center, Comet Tempel-Tuttle will be perturbed by Jupiter's gravity two years before its succeeding approach to perihelion in 2031. Its orbit will shift 2.4 million kilometers farther from the Earth's, the same thing that happened before the famous missing shower of 1899. So we won't be seeing a meteoric hurricane in 2031. It will be quite unlikely at the succeeding perihelion too—in 2065, when the separation will be 2.2 million kilometers. In fact, the many events in the past were produced only when the comet passed between 2.2 million kilometers inside and 1.5 million kilometers outside the Earth's orbit. Thus another century will pass before we see a Leonid storm pelt down again. After 1999–2000, it can happen again only in 2098, when the spread will narrow to 925,000 kilometers.

Remember that the Leonids are so spectacular because they are extremely fast, as the table on page 88 shows. Indeed, they possess exactly the maximum theoretical speed of an object colliding with our planet—72 kilometers per second—because, like their parent comet, they assume a retrograde motion around the Sun. Their collisions with the Earth's atmosphere, therefore, are almost head-on, giving rise to brilliant, fast, colorful (blue, green, white) meteors that leave trails lasting minutes at a time. Many Leonids observed just after the radiant rises cast ultralong paths across the sky.

Finding the Leonids' radiant is very easy. It lies inside Leo's unmistakable sickle-shaped head (see star map on page 111). This constellation can be found by using the Big Dipper. Draw a line between the last two stars of the Dipper, then extend the line seven or eight times its length away from Polaris.

Now draw a line between the Dipper's other two stars, and extend it onward to arrive at

Regulus, the brightest star in Leo. It would be futile to attempt this trick before midnight in mid-November, because Leo will not have risen completely. But around 2 a.m., the maneuver should prove quite handy. The Big Dipper will be low on the northeast horizon at that time, while Leo will sit just a little above the cardinal point east. The radiant will reach its maximum height just before dawn, around 5:30 a.m.

*Leonid trails seen at Greenwich, England, on the night of November 13/14, 1866. Location of the radiant is obvious.*

113

# Eclipses

*The path of an eclipse across the Earth's surface can be calculated millennia in advance.*

*Although it is entirely predictable, a total eclipse of the Sun is still perhaps the most sublime spectacle among the great cosmic phenomena.*

*In that brief time—the few moments of totality—years of anticipation are subsumed. It's a sight capable of stretching the most fervent imagination. And when the Sun at last reappears from behind the Moon's black shroud, something then makes you sigh with relief: "It's still there . . ."*

# Eclipses in History

## Pure Emotion

A total eclipse of the Sun is one of Nature's grandest and most impressive events, perhaps more imposing than any other. It is certainly difficult to compare it with the celestial phenomena we have already talked about. This, of all celestial events seen by the author, has been the most moving, perhaps precisely because of the briefness of the period of totality. Mine has been a case of emotion in its pure state, almost primordial—similar, I believe, to the sense of wonder felt by ancient peoples who stood in the grip of a total solar eclipse.

## The Law of the Sky

No wonder ancient peoples viewed eclipses as fateful signs—even more than the apparition of comets or meteor showers. Indeed, ancient civilizations had a profound relationship with everything that happened in the sky. They had to know how to extract fundamental signs from the heavens that could guarantee the survival of their communities, signs that reflected the procession of the seasons, atmospheric variations or the will of the gods. In effect, every celestial event had to have an earthly counterpart. No astronomical event was ever without consequences for the fate of the community.

Thus because life on Earth depended (and still depends today) upon the two great luminaries, the Sun and the Moon, the disappearance of one of these—especially the brighter one—even for a tiny period of time was considered an upheaval of cosmic magnitude. One such famous episode is recounted by Herodotus in his *Histories.* In the course of a rather long war between the Lydians and the Medes that had already lasted five years, ". . . while they engaged in battle during the sixth year, the day suddenly turned into night just when combat was the most heated. . . . When they saw night take the place of day, the Lydians and Medes quit fighting and, with deep solicitude, prayed that peace be made." The event cited is generally thought to be the eclipse of May 28, 585.

Usually, eclipses were viewed as the Sun or the Moon being eaten by some kind of monstrous creature. In China, it was a celestial dog; among South Americans, a bird with outstretched wings; in Armenia, a dragon. In India, the Sun and the Moon were being swallowed by Rahu,

*Preceding pages, sequence of images from the July 11, 1991, eclipse photographed over the cathedral at La Paz, in Baja California, by Akira Fujii, Kiroyuki Tomioka and Yonematsu Shiono.*

*Above, image of the Sun found in the Temple of Gold at Amritsar in India's Punjab region. (From L'Astronomia)*

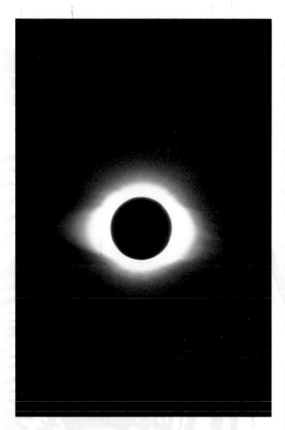

To exorcise the danger, the Buriats would emit hearty screams, cast stones and, more recently, fire weapons into the sky to scare the monster away. Other people employed—and sometimes still do—similar behavior. For example, the Kalina of Surinam made a hell of a noise and threw anything they could get their hands on to keep the two contenders apart, because even though they were brother and sister, the Sun and the Moon sometimes fought furiously.

In North America, members of the Ojibwa tribe living near the Great Lakes shot flaming arrows into the sky in an effort to reignite the Sun. The Sensos in Peru did the same thing, intending to scare the wild beast that was attacking the Sun. Ancient Colombians grabbed their

weapons and intoned hymns of war, promising the Sun and Moon that they would repent their sins and engage heart and soul from then on in minding their own business. And they would prove their zeal by dedicating themselves to irrigating the grainfields and working at a frantic pace during the event. During an eclipse of the Moon, medieval Germans would begin chanting, "Yay, Moon!"

In ancient Egypt, eclipses supposedly happened because a black pig, the demon Set in disguise, jumped into the eye of Horus, the Sun god. According to an ancient Norwegian tribe, the gods chained up the wizard Loki. To avenge himself, Loki created giant wolves, the biggest of which were Mânagarmer, who caused lunar eclipses by swallowing the Moon, and Sköll,

one of the demons that fought the gods for possession of the goddess Laksmi—the equivalent of Venus—and of a divine nectar similar to ambrosia. The demons won, but just when Rahu was drinking the ambrosia, the god Narayana surprised him and cut off his head. Ever since, Rahu has wandered the skies, lying in ambush for the Sun and the Moon.

Among the Buriats living east of Lake Baikal in southeastern Siberia, the belief in a monster named Alkha was widespread. Alkha continually chased and devoured the Sun and the Moon. He repeatedly darkened the world, until the exasperated gods cut him in two. The lower half fell to earth, but the upper half continues to stalk the skies, which explains why eclipses still happen every so often.

*Above, total solar eclipse of October 1995 photographed in India by Gianvittore Delaito.*

*Right, battle between the Sun and the Moon in a 14th-century codex. (From L'Astronomia)*

who caused solar eclipses by eating the Sun.

The Alaskan Inuit were convinced that solar eclipses were caused by the Sun's illness. In Transylvania, it was thought that humanity's perversion caused the phenomenon: to avoid seeing it, the Sun shuddered and turned away, hiding itself in disgust. A poisonous mist would then fall from the sky, the font of infection and epidemics that made it dangerous to stay outside and drink water or eat fresh fruits and vegetables. Other civilizations, instead, attached positive meanings to an eclipse, such as the belief in Bohemia that you could find gold during an eclipse.

Among many peoples, the phenomenon was seen as the union of two spouses. In Tahiti, it was very romantic: it was thought that the Sun and Moon could not find their way when darkness came, so they created the stars to relight their way. Among the Tlingit of Canada, the phenomenon was thought to occur when the Moon went to visit her husband, while the Algonquins believed it was due to the Sun taking his son into his arms. Inasmuch as Germans considered the Moon masculine and the Sun feminine, their traditional mythology explained an eclipse as the outcome of arguments between marital partners.

## Understanding the Phenomenon

Naturally, alongside the mythological interpretations, attempts to explain the physics of the eclipse phenomenon were quick to form, starting with the earliest significant explanation—that an eclipse was caused by an alignment of the Sun, the Moon and the Earth.

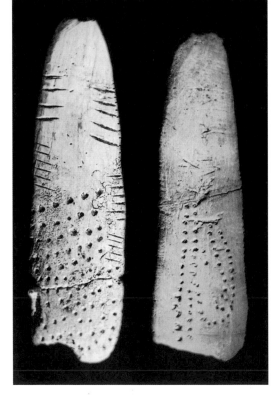

*Top, aerial view of the Stonehenge cromlech on Salisbury Plain in southern England. (A. Hasarth, Explorer)*

*Bottom left, image of Egyptian Sun god Athon found in tomb of Pharaoh Tutankhamen. (From L'Astronomia)*

*Bottom right, etchings on deer bone dating to the Early Paleolithic period found at Les Eyzies, France. According to American paleoanthropologist Alexander Marshack, they represent records of lunar activity. (Curcio Photographic Archive)*

Although there are no ancient enough testimonies in existence to prove it, we have reason to believe that Early Paleolithic humans were already able to record the phases of the Moon rather accurately and thus had a good starting point to approach the question. This seems to be demonstrated by the discovery, in Africa, France, the Ukraine and Spain, of animal bone fragments dating from 9,000 to 20,000 years ago that carry incisions arranged in sequences of 30. It is known that the great megalithic monument of Stonehenge, whose construction dates back almost 5,000 years, was an extraordinary lunisolar calendar, a device of alignment and orientation with the points at which the Sun and Moon would rise and set on various significant occasions. Among other things, there are certain numerical aspects that seem to have

something to do with the days in a lunar cycle. There are 30 stone arches in the configuration that makes up the so-called Sarsen Circle and 30 and 29 holes in two systems of cavities conventionally called Z holes and Y holes (the average of the two numbers, 29.5, corresponds to the number of days in the synodic lunar month, or the interval between two new Moons).

## Between Two Rivers

The first documents containing records of eclipses date to the eighth century B.C. and come from ancient Babylon. It's possible, of course, that Babylonian astronomers had already figured out how to record eclipses several centuries earlier. The famous *Gilgamesh Epic*, which dates to the First Sumerian Empire (around 2700 B.C.), constitutes an allegory of celestial movements related to the Sun and Moon. The Babylonians continued to record eclipses through the first century B.C. Their accounts are quite accurate. They measured the interval between the beginning of an eclipse and dawn or sunset, as well as the duration of the various phases, with the help of a water

clock and calculated the degree of totality of an eclipse as a fraction of the solar or lunar diameter. Time was counted in *us*, an interval of four minutes—the time required for the celestial sphere to appear to turn 1 degree due to the Earth's rotation.

A classical record compiled in 240 B.C., when the Seleucid dynasty had come to power in Babylon, sounds like this: "Eighth month, 14th day, at three *us* before dawn [a lunar eclipse] began on the eastern side [of the Moon. The Moon] set in eclipse." The Babylonian documents cover some 40 solar and lunar eclipses; only one record, however, concerns a total eclipse of the Sun, the one that happened on April 15, 136 B.C. This had to be a total eclipse because, according to the report, it became dark enough for the planets Mercury, Venus, Mars and Jupiter to become visible as well as "the normal stars."

Much has been written about the Babylonians' supposed ability to predict eclipses. Indeed, by the second millennium B.C., they probably knew about the so-called saros cycle (see Chapter 8)—that eclipse sequences repeat every 18 years and 11 days. This gave them a reliable method of predicting lunar eclipses, which are visible over an entire hemisphere. Predicting a solar eclipse in a given locality, however, would have required a more precise knowledge of the relative sizes and distances of the Sun and Moon.

Thus Babylonian astronomers could say only whether an eclipse was possible or not. Achieving a mathematical framework adequate enough to predict solar eclipses would have to wait for the work of Claudius Ptolemy, an astronomer who lived in Alexandria in the second century A.D.

*Top, the oldest written record of an eclipse is found on this Assyrian tablet from June 15, 763 B.C. (British Museum, London)*

*Bottom, symbol of the Sun on a Babylonian tablet from the ninth century B.C. (Curcio Photographic Archive)*

119

At about the same time Ptolemy flourished, the Mayan civilization began its expansion on the other side of the ocean. That population's astronomers discovered very early that eclipses could occur at intervals of five or six synodic lunar months. Thus even without the ability to predict an event, Mayan astronomers compiled charts that specified those (ill-starred) days when an eclipse was possible.

## Shadow of the Celestial Empire

In our earlier discussion of comets, we noted that eclipses were among the phenomena recorded in ancient Chinese annals. Curiously, we run into an anecdote here that unites history and myth. It tells the story of Hsi and Ho, court astronomers who lived around 2000 B.C., in the time of the legendary emperor Yao. One day, they were drunk. As a result, they were not up to observing an eclipse of the Sun, and for this, they were put to death. Although the episode is generally considered apocryphal, severe punishments were indeed meted out to astronomers who failed to keep their houses in order.

The first credible Chinese records of eclipses were written by the astronomers of a single little state called Lu and are more or less contemporaneous with the Babylonians' accounts dating to 720 B.C. At least 37 solar eclipses were recorded in Lu before 480, and three of them were total. The precision of the dates is remarkable, but unfortunately, an indication of the hour and any descriptive details are missing.

With the advent of the Han dynasty in 206 B.C., eclipse observations (and determining their astrological meaning) were transformed into an affair of state and therefore became much more accurate. This was also true for other celestial phenomena. Here is a descrip-

*Panoramic shot of the eclipsed Moon as it is about to set on September 27, 1996. Photograph by Corrado Marcolin.*

*Facing page, top, the Sun Rock exhibited at the National Museum of Anthropology in Mexico City celebrates the events of the Four Creations, or the four ages of the world, represented by the four so-called suns that sur-round the central figure of Tonatiuh, the Aztec Sun god. Photograph by the author.*

tion of a near-total solar eclipse that was recorded in 120 A.D.: "There was an eclipse of the Sun. It was almost complete, and it was almost as though evening arrived on Earth. [The Sun] was 11 degrees into *Hsü-nu* [the house of the Moon]. The sovereign displayed an aversion to all of this. Two years and three months later, Teng, the Emperor's mother, died."

The compendia of Chinese observations altogether contain about 10 accounts of total or annular solar eclipses. The account of 1292, in particular, is one of the few clear descriptions of an annular eclipse before the modern epoch: "The Sun looked like a ring of gold."

## From the Greeks to the Middle Ages

The Babylonians' descriptive and predictive mathematical apparatus in the field of astronomy passed to the Greeks, who contributed an elaboration of the first theories on planetary astronomy to the picture. Among other things, the Greeks coined the term "eclipse," from *ekleipsis*, which means "abandonment; to be flawed; to become less," undoubtedly referring to the aspect that the Sun or Moon assumes as the event progresses. Nonetheless, accounts of eclipses in ancient Greece are rather approximate. Ptolemy's *Almagest* is the best relevant source, but it contains references to only nine lunar eclipses.

During medieval times, observations of eclipses multiplied. In Europe, they were recorded—usually in a nontechnical way—in the chronicles of important events kept at various monastic centers. Typical of a monastic entry, somewhat confused but clear enough to understand that the event was not a total solar eclipse, is a record of the August 2, 1133, eclipse written in Prague: "An eclipse of the Sun

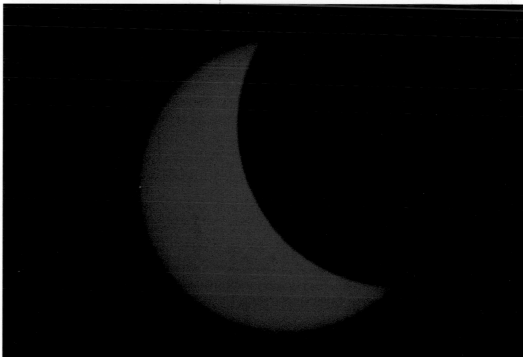

*Bottom, partial phase of the total solar eclipse of July 11, 1991, photographed by the author.*

121

appeared in a wonderful way. A flaw diminished it gradually to the point that a crown shape, similar to a crescent Moon, moved toward the southern part, rotating first toward the east and then moving west. At the end, it changed again, returning to its original state."

F. Richard Stephenson calculated that at least one-fifth of the historical accounts of eclipses

from every part of the world before the 1700s are found in the archives of Italian monasteries. The first credible record of a total eclipse seen in an Italian locale—Sicily—is much older, however, inasmuch as it refers to the event of August 15, 310 B.C. Found in the *Biblioteca storica* (*Historical Library*) of Diodorus Siculus (first century B.C.), it recounts the tyrant Agathocles' flight

*Above, lunar eclipse of February 9, 1990, photographed by the author.*

*Right, Chinese text found in Hou Han Shu (History of the Early Han Dynasty), which mentions the solar eclipses of 118 and 120 A.D. (Richard Stephenson)*

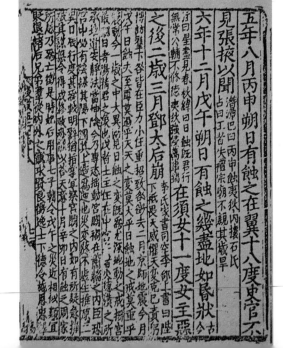

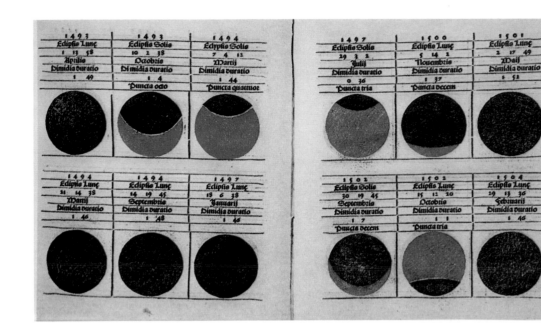

a table calculated for a place with a known longitude to get the difference in longitude between the site of observation and the site for which the table was calculated. Using this method, Arabian astronomer al-Biruni, active at Ghazna (now called Ghazni), in Afghanistan, around the year 1000, was able to calculate the difference in longitude between Ghazna and Gorgan, in Iran—1,300 kilometers apart—with an error of only 0.15 degree, equivalent to just 13.5 kilometers on the ground.

## Modern Uses of Ancient Eclipse Accounts

Ancient eclipse records are extremely important for at least two reasons. The first concerns

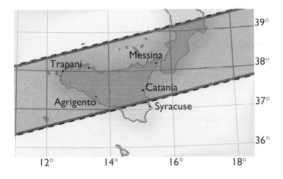

from Syracuse, which forced a Carthaginian naval blockade. According to Diodorus: "They were unexpectedly able to reach safety as dawn broke. The following day, there was an eclipse of the Sun, during which a thick darkness fell and stars were seen to shine all over the sky."

Taking a step forward in history, another Italian tale—this time written in medieval times about the eclipse of May 5, 840—again testifies to people experiencing great apprehension of these phenomena. The chronicler Andrea da Bergamo recounts: "During the third summons, the Sun was obscured and stars appeared in the sky on the third day before the ninth of May in the ninth hour of the litany of the Lord and for about a half-hour. There was great dread among the people, and many began to fear that this world of ours was coming to an end. But while they exchanged similar ingenuous thoughts, the Sun began to shine again, fleeing from the shadow that had completely enshrouded it earlier."

### Arabian Eclipses

In the Arabian nations that inherited the immense cultural legacy of classical Greece after 800 A.D., we find two types of eclipse records: the first executed with great accuracy by expert astronomers; the second casually annotated by various authors with a nontechnical approach.

In ancient Arabia, the observation of eclipses was important for two reasons: (1) a need to verify the accuracy of existing eclipse tables (it was believed that Arabian astronomers were able to predict a solar eclipse to within 40 minutes and a lunar one with even less disparity); and (2) the possibility of determining differences in longitude between locations on the Earth's surface. Lunar eclipses were used for this latter purpose. Since they happen simultaneously wherever on Earth they are visible, you need only record the exact time of the event in two locations to find the difference in longitude between the places or to compare the time of the eclipse in one locality with the time given by

*Above, phases of a total lunar eclipse reproduced in the Astronomicum Caesareum of Petrus Apianus. (Ingolstadt, 1568)*

*Top right, path of totality for the Sicilian eclipse of 310 B.C.*

*Bottom right, during the 15th century, various almanacs began to circulate featuring tables for predicting eclipses. In the illustration, a double-page spread from Johannes Regiomontanus' Calendarium shows lunar eclipses from 1493 to 1504. (Venice, 1485)*

the stability of the Sun's diameter. In 1979, John Eddy and Aram Boornazian advanced a hypothesis that the Sun is contracting. If this were true, total eclipses over time would last longer and longer, because the Moon would cover the Sun for a longer period. Since the amount of contraction would be very small, however, confirming the theory would require having very accurate records of the times of past eclipses, limiting the useful observations to only those made since the 17th century. The analysis thus far would seem to dispute the theory of contraction.

The second reason concerns the length of a day. We know that ocean tides, driven by the Moon, exert a braking effect on the Earth's rotation, thus stretching the day's duration. It has been calculated that in 100 years, the day's duration increases by about 45 seconds. During longer intervals, however, the difference grows greater as a function of the elapsed time. Thus 1,000 years ago, the disparity must have been 1.5 hours, and 2,000 years ago as much as five hours. By possessing records of eclipses that date back to those eras and beyond—however imprecise the notations—we can get sufficiently useful information on the trend of the Earth's rotational speed.

For example, if we failed to account for the changing length of the day, the path of totality for the March 4, 181 B.C., eclipse would not have crossed over Ch'ang-an (now Xi'an), China, the capital of the earlier Han dynasty, yet Chinese historical sources attest that it did. Or without the needed corrections, the path of totality of the 484 A.D. eclipse observed in Athens

*Top, the first photograph— or, rather, the first daguerreo- type—of a solar eclipse, taken July 28, 1851.*

*Bottom, telescopic view of prominences and chromo- sphere drawn by Lilian Martin-Leake during the eclipse of May 28, 1900.*

would have been displaced to the west by more than 20 degrees.

On the other hand, highly accurate records like the one for the eclipse of 136 B.C. demonstrate that the problem is more complex. Keeping in mind the simple slowdown related to the tides, the path of totality would have moved east of Babylon by more than 22 degrees. Evidently, other factors must tend to accelerate rather than slow the Earth's rotation, such as a decrease in the oceans' water levels, the contraction of the planet and the expansion of the Earth's core. The study of historical eclipses can help shed light on all these issues.

## Toward the Modern Era

It is possible that the first mention of the Sun's corona, the silvery halo of solar atmosphere that appears around the black disk of the Moon during totality, dates back to a Chinese inscription on oracular bones from 1307 B.C. It reads: "Three flames devoured the Sun, and a big star became visible." Yet this description could also refer to solar prominences. Another ancient citation is found in Plutarch's *De Facie in Orbae Lunae* (*Phases of the Moon*; first century A.D.), which speaks of a "glory" around the eclipsed Sun. The first certain observation comes from Corfu during the eclipse of 968.

In the early 1600s, Kepler was the first to propose that the corona belonged to the Sun, not to a hypothetical lunar atmosphere. Giacomo Filippo Maraldi of Italy was able to demonstrate this a few decades later by noting that the Moon crosses the corona during an eclipse, and the corona remains fixed around the Sun.

The term "corona" was probably used for the first time by José Joaquin de Ferre of Spain, who observed the eclipses of 1803 and 1806.

The first irrefutable observation of solar prominences is ascribed to Julius Firmicus Maternus during the July 17, 333, eclipse observed in Sicily.

Photography became available around the mid-1800s as a means to record eclipses. The

first image of a solar eclipse was a daguerreotype made during the July 28, 1851, eclipse by professional photographer Berkowski of the Königsberg Observatory in Prussia. The inner corona and prominences are visible. That same eclipse also offered conclusive proof that the prominences belong to the Sun, since the Moon's movement at the beginning of the event covered the prominences on the Sun's eastern rim and, at the end of the eclipse, uncovered those on the western rim.

During the eclipse of July 18, 1860, which was visible in Spain, England's Warren de la Rue and Italy's Angelo Secchi confirmed that the prominences belonged to the Sun when they were able to take far better photographs, thanks to the new wet collodion plates that permitted much shorter exposure times.

On the above map of the 136 A.D. eclipse, top, the central path (1) that passes through Babylon (2) corresponds to its description in a Babylonian astronomical record. The path at left (3) is the one the eclipse would have followed if the Earth's rotation had been the same then as it is now; the path at right (4), displaced 22 degrees to the east, shows where the path would have fallen if tidal effects were the only factor affecting the changing length of the day. A splendid drawing of the July 18, 1860, eclipse, bottom, executed by Wilhelm Tempel demonstrates how the human eye could be much more effective and versatile than the photography of that era in reproducing subtle details visible within the solar corona.

# The Eclipse Phenomenon

## Why Eclipses Happen

An eclipse of the Sun occurs when the Sun, the Moon and the Earth are lined up in space and the Moon (at new phase) lies between us and our star so that the Sun is partially or totally covered by the Moon. An eclipse of the Moon, on the other hand, happens when the Earth lies between the Sun and the Moon (at full phase) and the bulk of our planet keeps direct sunlight from hitting our satellite.

For an eclipse to happen when the Moon is in either of these phases also requires that our satellite be positioned near one of the two nodes of the Earth's orbit, those points where the orbits of the Moon and the Earth intersect. If these orbits were to lie in the same plane, there would be two eclipses every month—one solar and one lunar—at every new Moon and every full Moon. In fact, however, the Moon's orbit is inclined by just over 5 degrees from our planet's orbit.

An eclipse will be total or partial depending on whether the Moon is closer to or farther from the node at the time, making the alignment more or less exact. The regions that lie outside the center zone during a total eclipse of the Sun will have a partial eclipse, with less and less coverage of the Sun's disk as one gradually moves away from the path of totality.

Yet another form of solar eclipse can occur, called an annular eclipse. Since the Moon's orbit is elliptical, the Moon's distance from the Earth varies by almost 9 percent from perigee (minimum distance from Earth) to apogee (maximum distance). Consequently, its apparent diameter varies by the same amount, going from a maximum of 33'30" to a minimum of 29'22".

When the Moon is near apogee, its diameter is so small that even if it is perfectly in line with the Sun, it cannot hide the Sun's disk completely; a slender ring of Sun remains visible around the black disk of the Moon.

The apparent diameter of the Sun varies

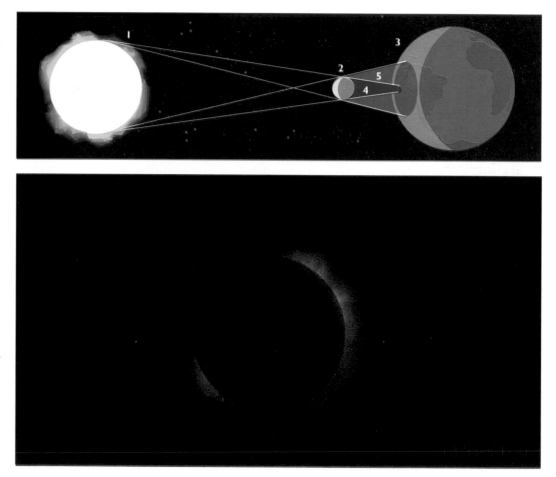

Top, geometry of a total eclipse of the Sun. The Moon (2), intercepting the rays of the Sun (1), produces a cone of shadow (4) and a penumbral cone (5) that reach the Earth (3). In the region immersed in the cone of shadow, there is a total eclipse; in the area within the penumbra, a partial eclipse.

Bottom, July 11, 1991, eclipse photographed by the author in Mexico.

Facing page, top, geometry of a partial eclipse of the Sun (1). The shadow cone (4) produced by the Moon (2)

does not reach Earth (3), which is touched only by the penumbra (5).

Center, eclipse of May 30, 1984, photographed by Carlo Zanandrea of the Feltre Astronomical Association Rheticus.

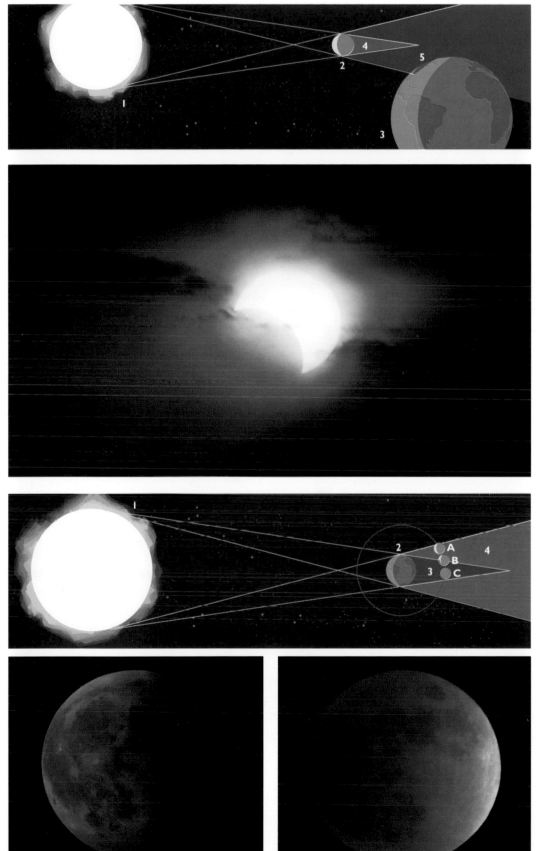

too—from 32'36" to 31'32"—since the Earth's orbit around our star is also elliptical. In winter, the Earth is 1 percent closer to the Sun than in summer. On average, the Moon's apparent diameter (31'05") is smaller than that of the Sun (31'59"), so annular eclipses are more frequent than total ones.

Among other things, the relationship between the apparent diameters of the Sun and the Moon determines the duration of each total solar eclipse. When the two diameters coincide, the phenomenon lasts for only a fraction of a second. The larger the Moon's apparent diameter, however, the longer the eclipse.

It is also possible to have an annular/total eclipse—an event seen as a total eclipse in certain parts of the world and as an annular eclipse in others, due to the curvature of the Earth. For regions where the Moon is high in the sky, our satellite is closer—even if ever so slightly—to the surface of the Earth and hence has a larger apparent size.

On the other hand, the maximum duration of an eclipse—theoretically, 7 minutes 31 seconds—can happen only when the Earth is near aphelion (the point of maximum distance from the Sun, around July 6 every year) and the Moon at perigee. This happens very rarely. The last eclipse longer than 7 minutes was on June 30, 1973. The next won't be until 2132.

## Space Geometry

That total eclipses of the Sun are possible seems an extraordinary coincidence. As it happens, even though the Moon is 400 times smaller than the Sun, it is exactly 400 times closer to Earth, which makes the apparent disks of the two great luminaries almost equal in size. Nature is full of coincidences, and in a certain

Center, geometry of a lunar eclipse. The Earth (2) lies between the Sun (1) and the Moon. If the Moon is completely within (C) the Earth's shadow cone (3), there is a total eclipse (see photo at bottom right, taken by author during Jan. 9, 1982, eclipse). If it is only partly in shadow (B), there is a partial eclipse (see photo at bottom left, taken by author during Feb. 20, 1989, eclipse). If it is inside the penumbral cone (4), the phenomenon will be imperceptible (A).

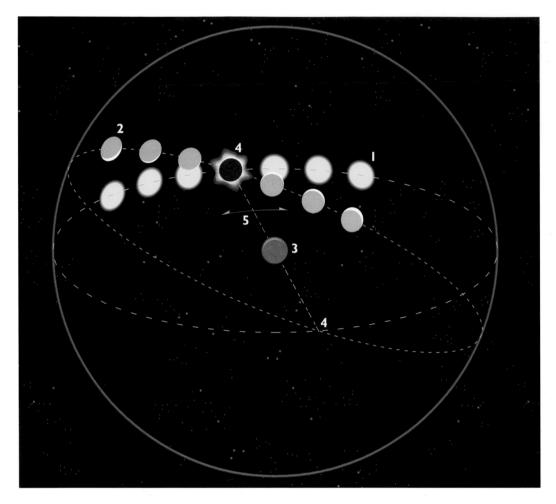

So eclipses happen during periods when the Sun reaches the node—twice a year, on average. In fact, although we typically express things in geocentric terms, the Sun's apparent movement through the year is actually just a reflection of the Earth's revolution around the Sun. Since the Earth intersects the Moon's orbit (reaches the node) twice on its annual journey, the Sun will, on those occasions, appear to be positioned in the sky at the opposite node.

The Sun accomplishes its journey between the nodes in 346.62 days, a period that we can call the eclipse year. It is shorter than the calendar year (the "tropical year"), which is 365.24 days, because the Moon's orbit is shifted in space due to the combined tidal forces of the Earth and the Sun. The line that unites the two nodes, therefore, is displaced toward the west by 19.4 degrees every year. As a result of this regression of the nodes, eclipses occur 18.62 days earlier each year in relation to the year before.

A minimum of two total or partial solar eclipses can occur every year, each time the Sun passes a node. But because of the width of the zone of probability, there will occasionally be four of them, possibly all partial. In fact, the possibility of a solar eclipse commences when the Sun is 15.35 degrees west of the lunar node and ends when it is 15.35 degrees east of the node. Our star moves about 1 degree per day (meaning, of course, its apparent motion as a result of the movement of the Earth around the Sun), so it takes 31 days to complete this course. Since the synodic month is 29.53 days long, the Sun cannot leave the zone of probability before the Moon arrives, and an eclipse should occur. Moreover, if the Sun finds the Moon at hand and an eclipse therefore occurs at the beginning of the 31-day period, it will

sense, the fact that things transpire is nothing more than the happy result of the meeting of various elements which were disunited before.

However, the question has another rather intriguing aspect. Because the Moon is slowly but continuously moving farther away from the Earth as a consequence of the tidal effect discussed in Chapter 7, total eclipses will eventually no longer occur. If the rate of the Moon's recession is maintained at about the same level as it has been over the last 400 million years— just under four centimeters a year—our satellite will no longer be able to hide the Sun in 600

million years. At least from this point of view, we live in fortunate times.

A solar eclipse can happen when the Sun's position is no more than 15.35 to 18.5 degrees from the node (the variation is due to the variability of the apparent solar and lunar diameters as well as the orbital velocities of the Earth and the Moon). If the distance is less than 10 or 12 degrees, the eclipse will be total or annular, depending on the Moon's distance from the Earth at that moment. A lunar eclipse can occur when the Moon's distance from the node is no more than 9.5 to 12 degrees.

*Above, apparent course through the sky of the Sun (1) and the Moon (2) seen from the Earth (3). At or very close to new Moon, a solar eclipse can occur only if our satellite is close to the node (4). The zone in shadow (5)*

*defines the limits within which an eclipse, whether partial or total, can happen.*

*Facing page, top left, geometry of an annular eclipse. Because the Moon (2) lies at a greater distance from the Earth (3), the Sun (1) will not be completely eclipsed, since the cone of the lunar shadow (4) does not reach the Earth.*

*In the area under the reverse cone, which we might call the antumbra, or antishadow (5), an annular eclipse will occur; while the region covered by the penumbra (6) will experience a partial eclipse.*

*Top right, annular eclipse of May 30, 1984, photographed by Loredano Ceccaroni from Sidi Moussa, Morocco.*

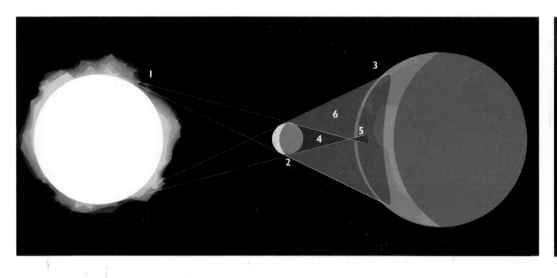

then be 30 days before it leaves the zone of probability. The result is that two partial eclipses can occur just one month apart.

Indeed, there could even be five solar eclipses in one year. Since the eclipse year is 19 days shorter than the tropical year, if an eclipse happens before January 18 (January 19 in a leap year), then there could be another one in January, two in July and one in December (belonging to the following eclipse year). This is a very rare event. It happened last in 1935 and, before that, in 1805, and it won't happen again until 2206.

In the case of lunar eclipses, since the Moon takes 19 days to cross the zone of probability for eclipses and this interval is less than the synodic month (or the time span between two full Moons), there is only one chance per nodal passage that a total or partial lunar eclipse can occur. Of course, there may be no eclipse at all. On the other hand, in the most optimistic scenario—when the first eclipse happens in January—there could be three lunar eclipses, total or partial, each year.

Altogether in one year, there can be a minimum of two eclipses (both solar) and a maximum of seven—three lunar and four solar or two lunar and five solar. Lunar penumbral eclipses are excluded from the count, as they are almost imperceptible.

## The Saros Cycle

The Babylonians discovered early on that eclipses of the Moon repeated every 18 years 11 days. This cycle, called the saros cycle, is easily explained by keeping in mind that it is a nearly exact multiple of several other cycles. In fact,

*Right, successive solar eclipses in the same saros series will not be seen from the same place but in localities 1, 2 and 3, respectively.*

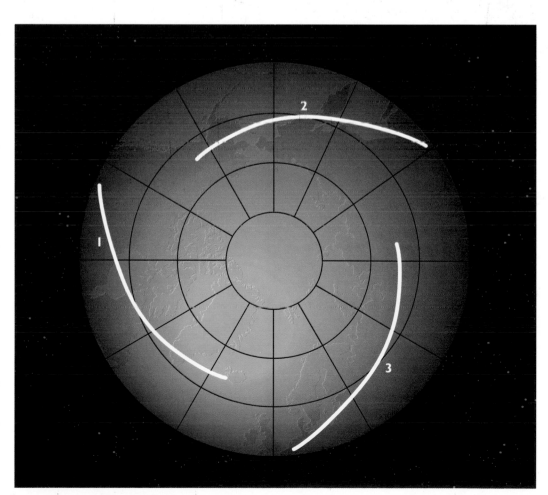

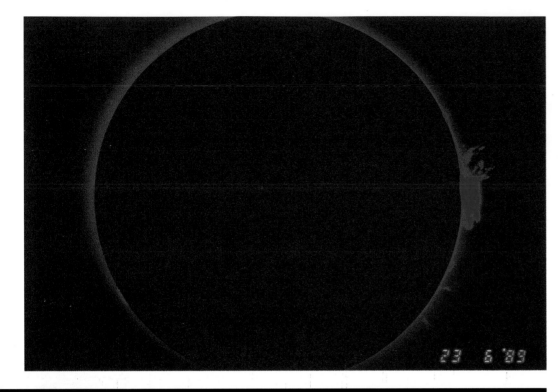

in 18 years 11 days, or 6,585 days, there are 19 eclipse years, 223 synodic months (return to the same lunar phase), 242 draconic months (return of the Moon to the same node; one draconic month = 27.21 days) and 239 anomalistic months (interval between two returns of the Moon to perigee, 27.55 days). The nearly perfect coincidence among all these cycles has the effect that at intervals of 6,585 days, more or less the same type of lunar eclipse occurs (with the same fraction of the satellite hidden), in the same season of the year, in the same part of the sky.

This is valid, in some measure, for solar eclipses as well, with the added fact that if the Moon was at the same distance from Earth in a total eclipse which happened 18 years earlier, you can be certain that the eclipse will be total again, rather than annular. Yet we must consider that the saros cycle is, strictly speaking, equivalent to 223 synodic months (29.53059 days x 223 = 6,586.3216 days) and 242 draconic months (27.21222 x 242 = 6,585.3572 days). Therefore, a saros equals exactly 6,585.34 days, or 18 years 11.3 days. This means that successive eclipses in the same saros will happen not at the same hour of the day but eight hours later (one-third of a day) and thus over an area

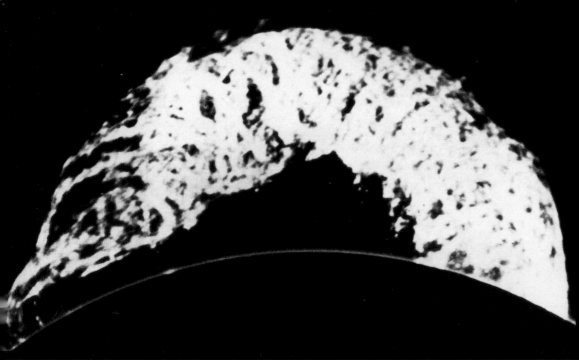

*Top, the chromosphere and prominences can be observed in the absence of an eclipse with an instrument called a coronagraph, which blocks the photosphere, and a special filter that permits only the red color predominantly emitted by these components to pass. The corona, however, is too faint to show up even with these expedients. Photograph by Claudio Bottari of Sava, Italy, with a 15-centimeter refractor telescope.*

*Bottom left, an enormous prominence rising 200,000 kilometers above the surface of the Sun. (HAO/NCAR)*

Facing page, bottom right, the chromosphere captured in Bolivia during the November 3, 1994, eclipse by Roberto Crippa and Cesare Guaita of the Tradatese Astronomy Group.

Above, bright prominence captured together with delicate coronal polar plumes during the July 31, 1981, eclipse in Kazakhstan. Photograph by Serge Koutchmy. (CNRS-IAP)

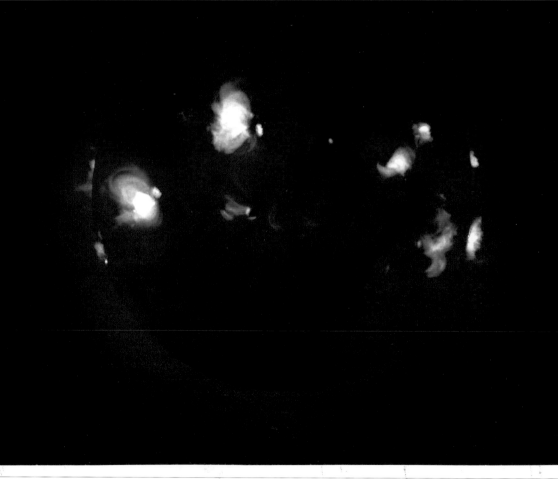

that is 120 degrees of longitude farther west. This does not preclude successive eclipses of the Moon from being visible from the same place, but it is exclusive for successive solar eclipses, which definitely will not be visible from the same site.

As you may have intuited, after three cycles, or 54 years 34 days, the eclipse returns to the same longitude. Because the eclipse does not happen in the same season, there will, however, be a displacement in latitude, whether north or south, of an average 1,000 kilometers, since the Sun will be at a different altitude above the horizon. The cycle of 54 years 34 days, already known to the Babylonians and the Greeks, was called *exeligmos*.

## Visibility

As you can see from the first illustration in this chapter, our satellite's shadow is very small at the distance of the Earth from the Moon. Due to the Earth's rotation, the cone of lunar shadow describes a path on the surface of the Earth a few thousand kilometers long. This is the eclipse's path of totality. The path is rather narrow. If the eclipse happens in tropical or temperate latitudes, the path rarely exceeds 300 kilometers in width. Yet when an eclipse takes place at a much higher latitude, whether north or south, the cone of the lunar shadow remains quite far from the equator of the Earth, and consequently, the shadow itself stays very close to the Earth's rim for the entire duration of the phenomenon. In the March 30, 2033, eclipse, for example, with the eclipse centered in northern Alaska, the maximum width will be a good 830 kilometers (with the Sun only 9 degrees above the horizon at that moment).

In any event, the narrowness of the path of

*Top, various features exhibited by the solar corona can be seen best with the aid of a differential-intensity filter (more opaque near the Sun's rim, where the corona is very bright, and more transparent farther out, where it is fainter). Photograph by Serge Koutchmy during July 31, 1981, eclipse. (CNRS-IAP)*
*1. Polar plumes*
*2. Coronal plumes*
*3. Coronal condensations*

totality explains why an eclipse is visible only in circumscribed zones. On average, any given locality will get one total eclipse every 410 years (Rome seems rather unlucky, with no total eclipses between 1567 and 2187). In the case of annular eclipses, the width of the zone of visibility is, as a rule, more or less the same, but here the exceptions are more notable. For example, on May 31, 2003, an annular eclipse centered between Iceland and Greenland will have a zone of visibility 5,185 kilometers in width (the Sun is only 0.5 degree high at that moment). Incidentally, the maximum duration of an annular eclipse is 12 minutes 30 seconds.

A total eclipse of the Moon, instead, has a remarkably long duration, since the Earth's shadow is truly immense, about 2.6 lunar diameters at the distance of the Moon. The Moon takes 3 hours 48 minutes to cross through the shadow centrally and proportionally less when it crosses it noncentrally (1 hour 44 minutes is the maximum theoretical duration for the phase of totality). Because of the Earth's great size, the entire hemisphere of our planet where the Moon lies above the horizon can witness the event, except where the eclipse occurs during the day. In practice, there is a 50 percent probability of witnessing a total eclipse of the Moon when it occurs. Since lunar eclipses happen once a year, on average, you can see one just about every two years.

## What Shows Up in the Dark

The Sun is a gigantic thermonuclear furnace that produces energy by transforming hydrogen into helium in its core at a temperature of 14 million degrees. After a very long journey, the particles that carry this energy emerge at the surface of the Sun, which is characterized by violent convective motions caused by intense magnetic fields and a differential global rotation, in which solar regions at different latitudes rotate at different speeds.

The interaction among these dynamic elements produces visible structures on the photosphere (the surface of the Sun that is normally visible when the Sun is not in eclipse and during phases of a partial eclipse). These include faculae (brighter zones where the magnetic field surfaces, which favors convection and therefore heating) and sunspots (zones where the magnetic field is so intense that it blocks convective motion and therefore cools the photosphere).

During an eclipse, some of the Sun's outer atmospheric regions can be seen. These regions are too faint to be perceived when the extremely bright photosphere is visible and are normally visible only with special instruments.

The chromosphere is the lowest, densest part of the solar atmosphere, and it circles the photosphere like a thin red hoop. Precisely because it is so thin, it is visible for only a few seconds at the beginning and end of totality, since at every other moment, it is hidden by the black disk of the Moon.

Prominences are enormous masses of extremely hot hydrogen that rise from the chromosphere tens or hundreds of thousands of kilometers into the corona. These can be considered coronal aberrations 100 times denser and 100 times colder than their surroundings. If it were not for the presence of the Sun's magnetic field, these would tend to dissolve into the corona within a short time, but instead, they last between a few hours (active prominences) and a few months (quiescent prominences). It is thought that prominences are formed by the condensation of coronal matter, in a fash-

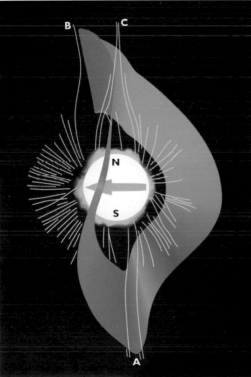

Facing page, bottom, the solar corona has an extremely high temperature, which makes it a potent emitter of x-rays. This x-ray image actually shows the corona across the entire solar disk because, among the Sun's various outer components, the corona is the only one whose temperatures are adequate to produce this type of radiation. The yellow spots are regions of greater emission heated to two million degrees by the action of the Sun's magnetic field. (Yohkoh Science Team)

Top right, splendid image of the corona obtained with a differential filter during the eclipse of February 16, 1980, which occurred at solar maximum. The most obvious features are the various coronal plumes (wide at the base and terminating in slender points far from the Sun) distributed almost uniformly around the entire disk of the Sun. (HAO/ NCAR, Rhodes College)

Bottom right, spatial structure of the outer solar corona, identified by the irregular green band. A, B and C indicate coronal plumes; the arrow on the Sun is the magnetic axis; N and S are the poles of the Sun's axis of rotation. The configuration is for the year 1999.

## 22 January 1992: Hα and white light

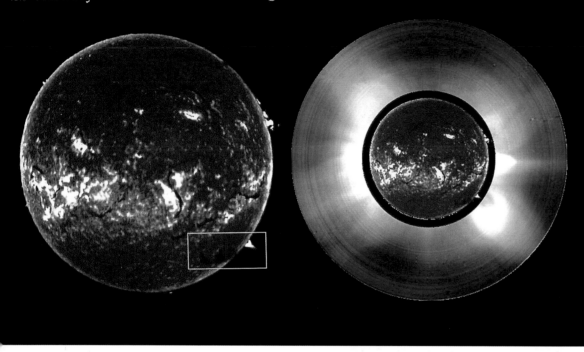

two solar diameters and is called the F corona due to the presence in its spectrum of the dark Fraunhofer lines that are caused by absorption by the coronal gas of light coming from the underlying solar layers. Its luminosity is produced by the scattering of sunlight when it encounters interplanetary dust. The third and faintest component is called the E, or emission-line, corona and is produced by light emitted by highly ionized atoms close to the Sun.

It is often said that the form of the solar corona varies according to the degree of solar activity, which follows an 11-year cycle, with characteristic minimum phases (few sunspots, small prominences) and maximum phases (lots of spots, remarkable prominences). It is generally accepted that during peak times, the corona looks predominantly symmetrical and circular, with brilliant radial extensions in every direction, while at times around the min-

ion similar to the formation of clouds in our atmosphere. A quiescent prominence averages 200,000 kilometers long, 50,000 kilometers high and 8,000 kilometers wide (the diameter of the Sun is 1.4 million kilometers).

The solar corona is the vast atmosphere of our star. Its luminosity is only one-millionth that of the photosphere. It looks like a blue-white halo around the Sun, irregular in shape, different from one eclipse to the next and varying in size between two and five solar diameters according to one's perspective. It is divided into three components. The first extends for about half a degree (one solar diameter) and is called the K corona (from the German *Kontinuum*), or electron corona, because its light is produced by the scattering of photospheric light by coronal electrons. The second extends as much as

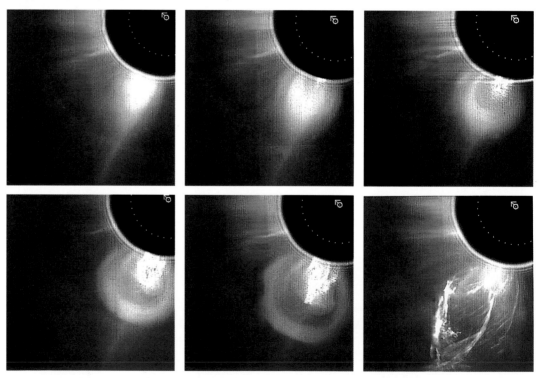

Top, coronal plumes are often superimposed on prominences. A prominence shown at left in red light (enclosed within a yellow rectangle) is revealed to be at the base of a coronal plume, at right. (Space Environment Laboratory, NOAA)

Bottom, sequence of six images documents the detachment of a coronal plume. This is one of the most dramatic manifestations of solar activity. In the case illustrated here, an impressive quantity of coronal matter—10 million

tons—rockets into space at a velocity of 2,000 kilometers per second. (HAO/NCAR)

Facing page, top, sequence of coronal images taken between 1966 and 1988 (four taken during total eclipses and one shot in space by the SMM satellite in 1985). Although the shape of the corona seems to correlate

with the solar maximums of 1968 and 1980 and the minimums of 1964, 1976 and 1987, it is primarily due to variations in the orientation of the outer coronal disk. (HAO/NCAR)

Facing page, bottom, eclipse of November 3, 1994, photographed by Serge Brunier in the Chilean Andes. In the foreground, the Parinacote and Pomerape volcanoes.

imum, it is concentrated in long plumes lying on the plane of the Sun's equator. Recently, however, a new interpretation of the corona's appearance imputes the changing look of the solar corona to a combination of solar rotation and the rotation—and inversion—of the Sun's magnetic field. According to this model, the solar corona is shaped like an irregular disk and is quite dense, while its plane is perpendicular to the Sun's magnetic axis. The magnetic axis, however, is not fixed in space but rotates in an arc of about 22 years around the Sun's axis of rotation. When our star is in the minimum phase, the magnetic axis is approximately perpendicular to the axis of rotation, while in the peak phase, the two axes are almost coincident. Halfway through the cycle, or about every 11 years, solar activity enters another minimum phase analogous to the one that happens every 22 years. It then has few spots and modest prominences, but its polarity is inverted with

relation to the preceding phase; that is to say, the Sun's north and south magnetic poles trade places. In effect, then, the real cycle of solar activity is 22 years, not 11.

Thus the corona does not actually become any more circular or symmetrical near the peak periods. This was formerly thought to be the case, because the corona is effectively brighter around solar maximum, making the corona's inner region appear overexposed and therefore partially concealing the long plumes in the outer regions. Only the use of differential-intensity filters and other techniques permits

a good view of the plumes in the peak phase. The same holds for the coronal polar plumes that dramatically display the lines of force in the Sun's magnetic field; our ability to view them was once considered possible only during minimum phases.

In conclusion, the appearance of the outer corona is primarily influenced by the Sun's rotation, which can totally change the geometry of observation in the sense that we will see the outer corona either in profile or face-on, and this is the only thing that governs the observed coronal symmetry or irregularity.

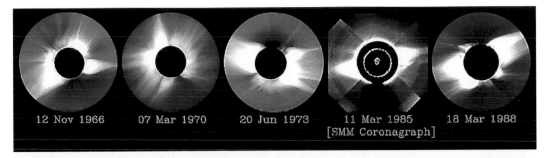

12 Nov 1966    07 Mar 1970    20 Jun 1973    11 Mar 1985    18 Mar 1988
[SMM Coronagraph]

# Lunar Eclipses

## When the Moon Hides

While it is logical that some part of the Sun, especially its atmosphere, would in theory remain visible during a total solar eclipse, considering the similar apparent sizes of the Sun and Moon, one would think that during a lunar eclipse, the enormous size of the Earth's shadow would conceal our satellite behind an impenetrable veil of darkness, making it literally disappear from the sky. In reality, the Moon's game of hide-and-seek is never perfect, which is what makes an eclipse of the Queen of the Night so tantalizing. The Moon never disappears completely from the sky. Sunlight is bent through the Earth's atmosphere and, thus deflected, makes its way into the Earth's shadow and strikes our satellite, giving it a deep vivid red hue. The color takes on a different shade from one eclipse to another, due to varying effects on its wavelengths sustained by the sunlight as it plows through the Earth's atmosphere.

## How Dark Is the Eclipse?

The amount of the Moon's disk that falls within the Earth's shadow during an eclipse is referred to as the "magnitude" of the eclipse. This plays a very important role in determining the brightness of the eclipsed Moon. Strictly speaking, the magnitude of the eclipse is the fraction of the lunar disk that is covered by shadow at maximum eclipse. A magnitude of 1 would

mean that the Earth's shadow exactly covered the Moon. A magnitude of 1.5 would mean that the Earth's shadow was half again the diameter of the Moon. Since the Earth's shadow at the distance of the Moon (384,000 kilometers) is 9,200 kilometers in diameter, or 2.65 times the Moon's diameter, it follows that the maximum magnitude of a dead-center lunar eclipse is 1.82 (the decimal part—0.82—is the area between the rim of the Moon and the rim of the Earth's shadow, expressed as a fraction of the Moon's diameter).

Now, if the eclipse is very central, sunlight will have great difficulty bending around the Earth to reach the Moon, and the Moon will tend to disappear. This happened, for example, with the eclipse of August 17, 1989 (magnitude 1.605). Other times, however, the level of atmospheric pollution on Earth is a factor. During

the eclipse of December 9, 1992, the Moon took on a very dark, almost black-looking red color due to dust from the Pinatubo volcano in the Philippines, which had erupted the year before.

In any event, the eclipsed Moon always experiences a very distinct drop in luminosity. The full Moon normally has a magnitude of –12.7, which is 2,000 times that of Venus. During a relatively bright total eclipse, the Moon shines roughly as brightly as Jupiter (magnitude –3), thus dropping in brightness by a factor of almost 120,000. Very dark eclipses, on the other hand, dim our satellite to around magnitude 4 (as in the eclipse of December 30, 1963, when our atmosphere was blackened by dust emitted from the Agung volcano in Bali), a decrease in brightness of at least five million times. Under those conditions, it becomes very hard to see the Moon with the naked eye. As we have noted

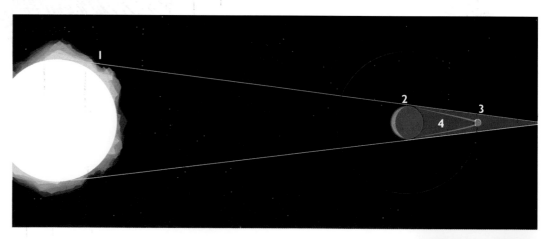

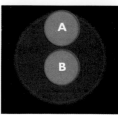

*Above, during a total lunar eclipse, light from the Sun (1) is deflected by the Earth's atmosphere (2) so that it reaches the Moon (3) even while our satellite is immersed in the cone of the Earth's shadow (4).*

*Right, relative sizes of the Moon's disk and the Earth's shadow. At A, the magnitude of the eclipse is 1 (Moon just inside the Earth's shadow); at B, it is 1.58 (Moon's rim is 0.58 of a lunar diameter from edge of shadow).*

previously, the naked eye can see stars to sixth magnitude, but the Moon's broad dimensions cause its light to be scattered across a larger area, giving the impression of less brightness.

Incidentally, both the 1989 and 1992 eclipses were estimated at a brightness of magnitude 3. Curiously, the magnitude of an eclipse can even vary from one observing location to another. This was true, for example, in 1989, when the eclipsed Moon appeared very dark in Europe but completely normal in North America. Since the Moon's height above the horizon can also play a role (in 1989, for example, it was a lot higher over North America than when seen in Italy), local atmospheric conditions are an important factor (a more or less clear sky, the presence of clouds in surrounding areas, the effects of atmospheric pollutants, etc.).

Top, sequential image of August 17, 1989, eclipse photographed by Carlo Ferrigno, Gabriele Vanin and Carlo Zanandrea.

Bottom left, the eclipse of December 9, 1992, one of the darkest ever seen, captured by Claudio Bottari.

Bottom right, April 4, 1996, eclipse, shot with the author's telescope.

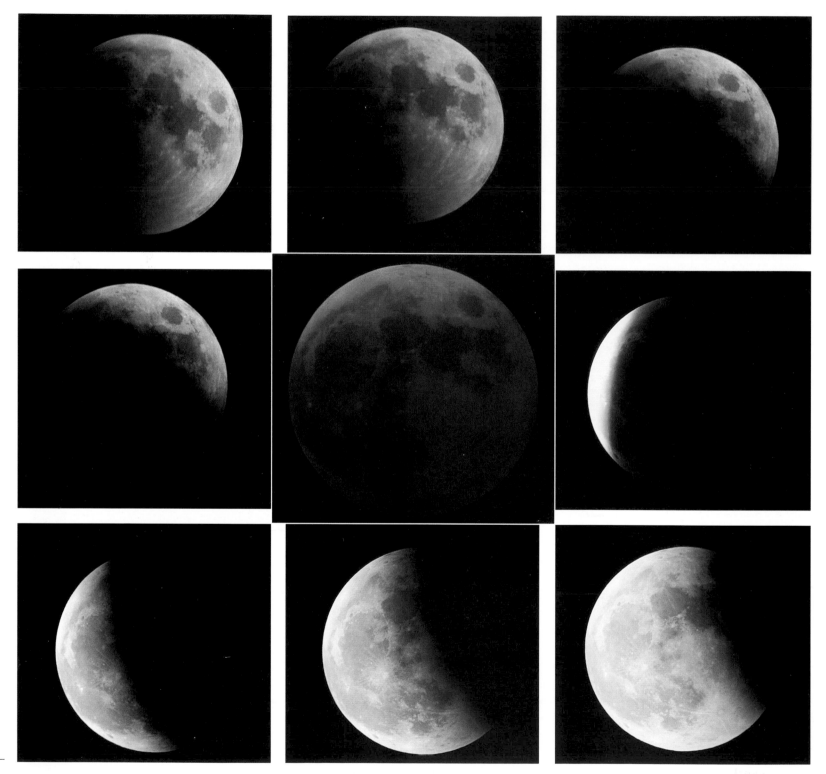

138

*A spectacular series of images showing the progress of the lunar eclipse of October 17, 1986, as the Moon enters and leaves the Earth's shadow, taken by Carlo Zanandrea at the Feltre Observatory.*

In some other cases, a clear difference in illumination between one sector of the Moon and the other can be seen. For example, during the November 19, 1993, eclipse, observers had the impression that the lower part of the Moon was never in shadow, while the upper part was very dark, an effect explained in part by the fact that the magnitude of the eclipse was just 1.09, so the Moon's lower rim was only 300 kilometers from the edge of the shadow.

The colors of the eclipsed Moon can range from a blend of reds and oranges all the way to yellow, the latter generally indicating greater luminosity. The April 1996 eclipse, for example, was very bright and appeared to be dark yellow when observed with binoculars or a telescope. Usually, the darker colors show up in lunar regions farther from the edge of the Earth's shadow and the brightest ones in regions closer to the shadow's edge.

As totality gradually approaches, there is always a show within the show. Prior to the eclipse, with the full Moon illuminating the sky, only the brightest stars can be seen. The sky then darkens progressively until stars that are normally visible at new Moon can be seen and—if the hour and the season are favorable (especially in summer months)—even the enchantment of the Milky Way.

**Observing Lunar Eclipses**

A partial eclipse of the Moon generally does not merit special attention, so we will concentrate exclusively on total eclipses. Every total eclipse of the Moon visible in North America from 2000 to 2010 is listed in the table on page 143.

Eclipses of the Sun and Moon repeat at an interval of almost exactly 19 years (the periodicity is especially good, but not perfect, for lunar

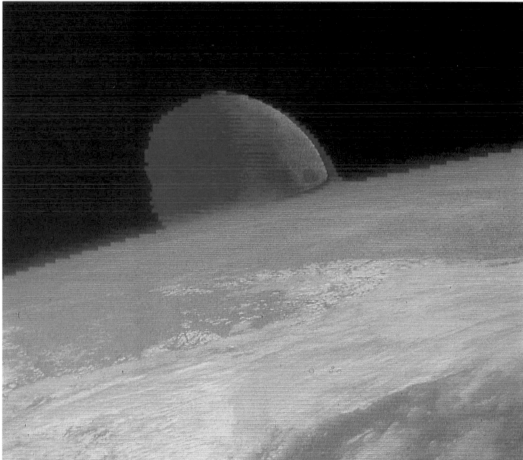

*Top, February 9, 1990, eclipse photographed by Carlo Zanandrea with Feltre Observatory's 20-centimeter Newtonian telescope.*

*Bottom, the eclipse of October 17, 1986 (see page 138), is seen here in an exceptional image taken in space by a geosynchronous satellite at 33,000 kilometers. Courtesy of Jay Anderson, Environment Canada.*

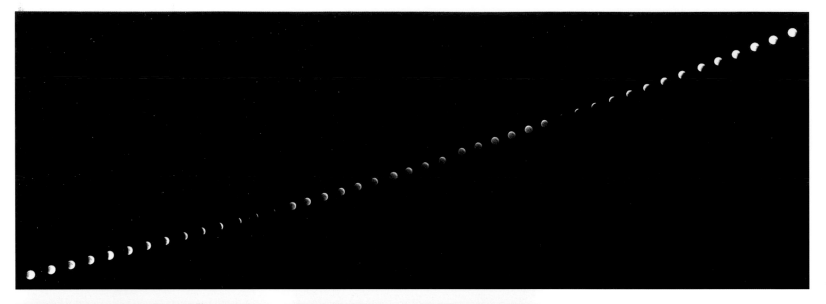

eclipses, although we are often dealing with phenomena of different magnitudes from one to the next and, in the case of solar eclipses, of different types that are visible in different places). This is the so-called Metonic cycle discovered by the eponymous Athenian astronomer in the fifth century B.C. Since 235 synodic months are exactly equivalent to 19 tropical years, the same lunar phases repeat at an interval of exactly 19 years. If we observe that this interval of time almost exactly equals 225 draconic months—the time required for the Moon's journey to the same node—we explain why eclipses repeat on this cycle.

The naked eye alone can usually see all the phenomena that accompany an eclipse—the progressive darkening of the sky, the appearance of bright stars and constellations around the Moon and the perception of the Moon's more or less attenuated brightness. But the event is definitely more enjoyable with binoculars. Colors are perceived better in binoculars,

*Top, sequential image of October 17, 1986, eclipse taken by the author with a 35mm lens and 50 ISO film.*

*Bottom left, the Moon and Saturn above a city street during the September 27, 1996, eclipse. Photograph by Corrado Marcolin with a 50mm lens.*

*Bottom right, another shot of the September 27, 1996, eclipse dramatically framed by Corrado Marcolin with a 135mm lens.*

and the differing shades and illumination of various parts of the satellite are easier to distinguish. A telescope generally adds nothing to the aesthetic pleasure of this kind of observation; in fact, it is counterproductive, because its excessive magnification dilutes the Moon's brightness and the perception of its colors. We can say, without reservation, that binoculars, especially if mounted on a tripod, are the best instruments for observing a total eclipse of the Moon. A pair of 20x80 binoculars are ideal, but even smaller types, like the classic 7x50 size, perform remarkably well.

## Photographing Eclipses

Capturing an eclipse photographically presents little difficulty, even without buying an expensive telephoto lens. Many people have used an ordinary camera with a normal lens mounted on a tripod. One of the most dramatic types of photography is the so-called sequential image, a series of shots taken at appropriate intervals

*Above, December 9 eclipse photographed by Claudio Costa at Gandolfo Castle. Owing to the extreme darkness of the event, a 2-minute exposure with a 135mm lens and 200 ISO film revealed many stars around the Moon.*

*Above right, September 27, 1996, eclipsed Moon over Port'Oria and the Feltre castle photographed by the author with a 135mm lens.*

(usually five minutes apart) that exploit the Moon's movement across the sky due to the Earth's rotation and reproduce the entire event on just one frame of film. It starts when the Moon enters the Earth's shadow and stops when it exits. In theory, this photograph would require a camera that permits a double exposure; that is, one which allows the shutter to open repeatedly without advancing the film. Few cameras are built with this capability, but you can work around it.

First of all, you must choose a lens that guarantees catching the entire course of the eclipse.

Since the 50mm lens is usually inadequate, it is better to opt for a 35mm and avoid surprises. Then, at the beginning of the eclipse, position the Moon in the lower left corner of the viewfinder and orient the camera on the tripod so that the diagonal of the frame is positioned along the course that the Moon will follow in the sky (observe the Moon's motion before the eclipse begins).

At this point, cover the lens with a lens cap and lock the shutter open (on the "B" setting) after having stopped the diaphragm to its smallest aperture (usually f/16 or f/22). The shutter stays open through to the end of the eclipse, but the lens cap keeps light from filtering in. Every five minutes, remove and replace the lens cap as quickly as possible, being careful not to shake the whole apparatus. With this kind of manual shutter release, you expose every shot for about 1/4 second. Use a normal 50 ISO film. When the partial phases of the eclipse begin, you will obtain a few images that are slightly overexposed (Moon a little too bright) but completely acceptable. Going forward, with the Moon obscured 25 percent, the 1/4-second exposure time will work very well. When the amount of Moon hidden increases to 50 percent, you can execute the movement a little more slowly (1/2 second), and when three-quarters of the Moon is within the Earth's shadow, increase the exposure to 1 second. When totality is about to begin, a few seconds' exposure will be good.

To shoot totality, you will have to perform differently, since the available light is very low. As soon as the cap is removed for the first shot during totality, open the diaphragm completely (this is usually f/2.8 for a 35mm lens). Practice first and work with extreme caution. You now have over 30 times as much light entering as

before, and 10 seconds will suffice to get a correct exposure of the eclipsed lunar disk (expose for 20 seconds if the eclipse seems unusually dark—for example, if the Moon is hard to see with the naked eye). At the end of totality, decrease the lens aperture again and follow the same steps in reverse until the eclipse ends.

Panoramic images of totality that include a landscape or an architectural element can be quite memorable, as long as they are moderately illuminated. Excessive illumination bleaches out the frame in an annoying way, and under inadequate light, the surrounding elements are barely visible.

You can use lenses from 35mm to 200mm in length, according to your preference and the equipment available. With ordinary film of about 100 ISO and a lens at f/2.8, the basic exposure time is roughly 5 seconds (10, if the eclipse seems dark). Remember that you can expose for up to 20 to 30 seconds with a 35mm

or 50mm lens without appreciable risk of the images being blurred by the Earth's rotation. Using a 135mm or 200mm lens, however, you must not exceed 10 and 17 seconds, respectively. Understand that the 20-to-30-second exposures with the shorter lenses are reserved for only the darkest eclipses photographed far away from city lights. This is also a way to capture a multitude of stars in the frame.

Without attaching the camera to a telescope with a clock drive that compensates for the Earth's rotation, the most a neophyte can do with a fixed camera is to shoot totality with a 500mm lens. Unfortunately, such lenses are very low-light by definition (usually f/8), so you must use a more sensitive film to get an acceptable image without risking camera shake. Using an emulsion of 800 to 1000 ISO, for example, and a 5-second exposure will produce a large enough image of the Moon with a level of vibration that is still tolerable.

### TOTAL ECLIPSES OF THE MOON VISIBLE IN NORTH AMERICA FROM 2000 TO 2010

| Date | Enters Shadow | Totality Begins | Maximum Eclipse | Totality Ends | Leaves Shadow | Magnitude |
|---|---|---|---|---|---|---|
| Jan. 20, 2000 | 22:01, EST | 23:04, EST | 23:43, EST | 00:22, Jan. 21 | 01:25, Jan. 21 | 1.33 |
| July 16, 2000[1] | 04:58, PDT | 06:02, PDT | 06:56, PDT | 7:50, PDT | 8:54, PDT | 1.77 |
| Jan. 9, 2001[2] | 13:42, EST | 14:50, EST | 15:21, EST | 15:52, EST | 17:00, EST | 1.20 |
| May 15, 2003[3] | 22:03, EDT | 23:14, EDT | 23:40, EDT | 00:06, May 16 | 01:17, May 16 | 1.13 |
| Nov. 8, 2003[4] | 18:32, EST | 20:06, EST | 20:18, EST | 20:30, EST | 22:04, EST | 1.02 |
| Oct. 27, 2004 | 21:14, EDT | 22:23, EDT | 23:04, EDT | 23:45, EDT | 00:54, Oct. 28 | 1.31 |
| March 3, 2007[5] | 16:30, EST | 17:44, EST | 18:21, EST | 18:58, EST | 20:12, EST | 1.24 |
| Aug. 28, 2007[6] | 01:51, PDT | 02:52, PDT | 03:37, PDT | 04:22, PDT | 05:23, PDT | 1.48 |
| Feb. 20, 2008 | 20:43, EST | 22:01, EST | 22:26, EST | 22:51, EST | 00:09, Feb. 21 | 1.11 |
| Dec. 21, 2010 | 01:32, EST | 02:40, EST | 03:17, EST | 03:54, EST | 05:02, EST | 1.26 |

Facing page, top left, the Moon during the February 19, 1990, eclipse photographed with a 500mm lens by Carlo Ferrigno and Davide De Bortoli of the Feltre Astronomical Association Rheticus.

Top right, October 28, 1985, event captured by Carlo Zanandrea with Feltre Observatory's 20-centimeter telescope.

Bottom, September 27, 1996, eclipse photographed through the author's telescope.

Above, table listing total lunar eclipses visible in the United States and Canada for the coming decade. Note that you may need to convert times according to your local time zone.

[1] Total in Hawaii; partial for West Coast of North America.
[2] End of totality visible from Atlantic Canada; partial in eastern Canada and northeastern U.S.
[3] Not visible in far northwest.

[4] Not visible on West Coast.
[5] Total for Great Lakes region and eastward; partial for most of continent.
[6] Total for Rocky Mountains region and westward; partial for most of continent.

# Solar Eclipses

### Partial Eclipses

A partial eclipse is not even remotely comparable to the excitement, splendor and scientific interest of a total eclipse of the Sun. In a total eclipse, the shadows fall and a unique series of events transpire, but more about this later.

A partial eclipse of the Sun is not as rare as a total eclipse. On average, in a given location, you can expect to see one every two or three years. Of course, we're talking about statistical averages—long intervals can sometimes pass with either no eclipse at all or several eclipses. In Rome, for example, not one solar eclipse was seen between 1984 and 1994, but there will be at least three between 2003 and 2006. More-

over, an eclipse seen as partial in one place can sometimes be annular or total in another or partial with less or more magnitude.

Since a partial eclipse of the Sun is an interesting phenomenon from an astronomical perspective and since not everyone has a chance to travel to witness a total eclipse, we have compiled a table on page 145 that lists every partial eclipse visible in North America through 2005.

### Annular Eclipses

An annular eclipse is also less spectacular than a total eclipse. Still, unlike a partial eclipse, it does produce significant observable environmental effects, such as an appreciable darken-

ing of the landscape. It is generally true that no great reduction of light will be noticed until an eclipse exceeds a magnitude of 0.5. When it reaches 0.8, however, the darkening is truly noticeable, and it begins to mute the color of the landscape. This occurs when only a sliver of Sun close to its rim remains visible. When you focus on a central part of the solar disk, you are peering through a relatively small portion of the Sun's gaseous envelope, enabling you to scan farther toward the center and to see the hotter, brighter layers. Conversely, when you look at the rim of the Sun, the density of its atmosphere is greater, so you cannot penetrate as deeply; you can discern only the coldest and therefore

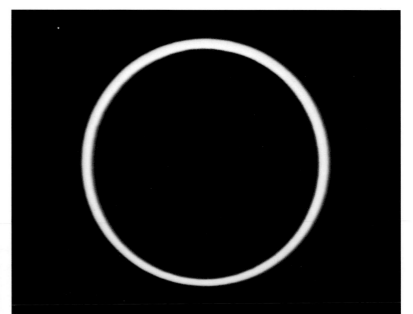

*Left, partial eclipse of October 12, 1996, photographed by the author from Feltre Observatory at moment of maximum magnitude (0.59).*

*Above, annular eclipse of May 10, 1994, photographed*

*in El Paso, Texas, by Loredano Ceccaroni of Rome.*

*Right, relationship between eclipse magnitude and percentage of the Sun's disk that is covered.*

| Relationship between eclipse magnitude & percent covered | |
|---|---|
| Mag. | Covered |
| 0.1 | 4% |
| 0.2 | 10% |
| 0.3 | 19% |
| 0.4 | 28% |
| 0.5 | 39% |
| 0.6 | 50% |
| 0.7 | 62% |
| 0.8 | 75% |
| 0.9 | 87% |

darker layers. The Sun's rim thus appears heavily obscured and redder than the rest of the disk. When the magnitude of an eclipse exceeds 0.8, the level of illumination falls off so quickly that it seems to herald imminent disaster. The Sun's radiance during annularity often diminishes as though a cloud were passing in front of it. During the May 10, 1994, eclipse in North America, observers said that the light diminished enough around midday to simulate late afternoon. In general, the sky becomes just dark enough to make out Venus and sometimes Jupiter but no other planets or stars.

When the rims of the solar and lunar disks coincide at the beginning and end of annularity (second and third contacts), you can see Baily's beads—points of light produced when the solar photosphere peeks between the mountains and valleys of the Moon. These are easier to see with annular eclipses of greater magnitude (varying, theoretically, from 0.901 to 0.999). Photographs of the chromosphere, prominences and even the whole corona have been published, but as far as we know, no one has ever perceived these details visually during an annular eclipse.

Unfortunately, annular eclipses are almost as infrequent as total ones and are far less interesting. In the end, however, we believe that it is entirely worthwhile to make the effort to observe an annular eclipse when there is one in your area. Details about annular eclipses that will be visible in Europe, Asia or Africa through 2020 are given in the map on page 146 and the table on page 147. In North America, an annular eclipse will be visible in parts of the western United States on May 20, 2012, and on October 14, 2023. Canada's Far North and northern Ontario will be in the path of an annular eclipse on June 10, 2021.

| Partial Solar Eclipses Visible in North America Through 2005 | |
|---|---|
| Date | Portion of continent where eclipse is visible |
| Dec. 25, 2000 | Northeast |
| Dec. 14, 2001 | West |
| June 10, 2002 | West |
| April 8, 2005 | South of line from San Diego to Philadelphia |

*Eclipse times will vary widely depending on the observer's exact location. Check almanacs, astronomy handbooks or astronomy magazines in the given year for details.*

| Partial Solar Eclipses Visible in Italy Through 2010 | | | | | |
|---|---|---|---|---|---|
| Date | City | Begin | Max. | End | Mag. |
| May 31, 2003 | Venice | (03:19) | (04:18) | 05:17 | 0.82 |
| | Turin | (03:23) | (04:21) | 05:19 | 0.83 |
| | Rome | (03:15) | (04:13) | 05:11 | 0.79 |
| | Bari | (03:09) | (04:08) | 05:08 | 0.77 |
| | Palermo | (03:09) | (04:07) | 05:05 | 0.75 |
| Oct. 3, 2005 | Venice | 08:49 | 10:14 | 11:38 | 0.65 |
| | Turin | 08:42 | 10:08 | 11:34 | 0.72 |
| | Rome | 08:47 | 10:16 | 11:44 | 0.72 |
| | Bari | 08:55 | 10:23 | 11:50 | 0.67 |
| | Palermo | 08:48 | 10:20 | 11:52 | 0.79 |
| Mar. 29, 2006 | Venice | 10:37 | 11:42 | 12:47 | 0.54 |
| | Turin | 10:34 | 11:36 | 12:37 | 0.48 |
| | Rome | 10:30 | 11:38 | 12:46 | 0.60 |
| | Bari | 10:31 | 11:43 | 12:54 | 0.69 |
| | Palermo | 10:22 | 11:34 | 12:46 | 0.68 |
| Aug. 1, 2000 | Venice | 10:10 | 10:43 | 11:16 | 0.09 |
| | Turin | 10:10 | 10:35 | 10:59 | 0.05 |
| | Rome | | not visible | | |
| | Bari | | not visible | | |
| | Palermo | | not visible | | |
| Jan. 15, 2010 | Venice | (06:21) | (07:07) | 07:54 | 0.16 |
| | Turin | | not visible | | |
| | Rome | (06:07) | (07:00) | 07:53 | 0.20 |
| | Bari | (06:10) | (07:02) | 07:54 | 0.18 |
| | Palermo | (05:53) | (06:52) | 07:52 | 0.25 |

*Left, partial solar eclipses visible in North America up to 2005 and in Italy through 2010 (times in parentheses indicate the Sun has not yet risen; times are Central European standard time).*

*Above, instant of second contact during annular eclipse of January 5, 1992, shot at Tabiteuea atoll in the Gilbert Islands. Photograph by Giancarlo Gengaroli.*

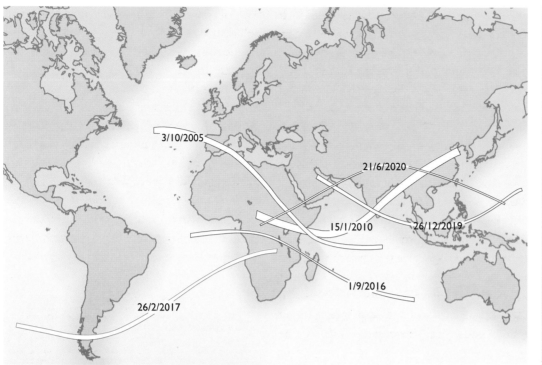

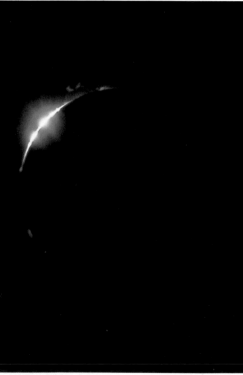

## Totality

There are four especially notable moments during a total eclipse: first contact, when the Moon begins to notch the solar disk; second contact, when totality begins; third contact, when totality ends; and fourth contact, when the Moon definitely deserts the solar disk. Every important phenomenon during the event, however, takes place around second and third contacts. In these moments, we can observe Baily's beads (named for English astronomer Francis Baily, who first wrote about them after the annular eclipse of 1836), the chromosphere, the prominences and the corona, as well as shadow bands, the diamond ring, the eclipse wind and the advance of the Moon's shadow across the landscape.

The diamond ring is a phenomenon produced when a bright Baily's bead (the diamond) flashes into view with a sliver of the inner corona. It is often spectacular to the naked eye. The author had occasion to observe one at the end of totality during both the 1994 eclipse in Peru and the 1995 eclipse in India. In the first case, the phenomenon was also tinted with a marvelous crimson hue, due to the reflection of chromospheric light by thin clouds that were present; in the second, the phenomenon seemed to last an eternity but was really only four or five seconds.

The eclipse wind is a cold wind generated by the temperature drop that follows the sudden disappearance of the Sun and the consequent cooling of the atmosphere. Sometimes

*Top left, bands of visibility for six annular eclipses cataloged in table on page 147.*

*Top right, during the December 22, 1870, eclipse, several observers in Sicily saw shadow bands move rapidly across the wall of a house.*

*Bottom, Baily's beads, the diamond-ring effect, the chromosphere and prominences are all captured in this splendid image by Roberto Crippa and Cesare Guaita taken during the 1991 eclipse in Mexico.*

*Facing page, table listing annular eclipses visible in Europe, Asia or Africa through 2020. Date is in the first column, duration of event in the second, width of the band of annularity in the third, magnitude in the fourth, Sun's alti-*

*tude in the fifth, local time of maximum eclipse in the sixth and, in the seventh column, the country that is most easily reached by people living in Italian locations or is most interesting as a tourist destination (in boldface type).*

*Facing page, bottom, testimony to the extreme darkness produced by the July 31, 1981, eclipse is furnished by this wide-field image taken by Stephen Edberg aboard a high-flying airplane.*

the opposite effect occurs, as in Mexico in 1991; the wind is felt first but falls off suddenly just moments before totality.

The approach of the Moon's shadow over the ground, starting about 15 seconds before the beginning of totality, has been observed during many eclipses. This is surely one of the more disturbing manifestations accompanying totality, and in the most dramatic cases, it is like a "dark and threatening storm," a "black monolith" that advances at hurricane speed, as described by Angelo Secchi during the eclipse of 1860. It is not always visible, however; on the contrary, we would say it is not easily seen.

Shadow bands seen during an eclipse are peculiar light and dark stripes that seem to play tag as they run across the ground. They are often seen a few minutes before second contact. At first, they appear as curious plays of light, chiaroscuros, like the evanescent luminous dappling when sunlight passes over forest leaves. They are caused by a phenomenon analogous to stars twinkling at night. To be precise, they are produced by turbulence in the layers of atmosphere a few meters off the ground. As totality approaches, the temperature drops significantly. This generates convection currents that modify the temperature and density of various layers of air, which in turn alters their index of refraction. Thus every little gust of wind causes a new deflection of light, and this produces the phenomenon. Then, as totality approaches, the dappled spots become discrete bands spaced a few centimeters apart. This banded profile is produced by the image of the crescent Sun projected on the ground (which gets progressively thinner in the seconds immediately preceding second contact) after passing through the turbulence present in the highest layers of the atmosphere. When totality ends, this phenomenon follows a reverse profile.

The author has had only one opportunity to witness this phenomenon, during the 1995 eclipse in India, two minutes before second contact and three minutes after third contact.

This is rather strange, because given what Johanan Codona—one of the foremost experts on the subject—has said, shadow bands should be easy to see in eclipses of long duration and very hard to see in short ones. Instead, we could not see anything in two other eclipses we witnessed, one of which lasted seven minutes and the other three minutes, while the Indian eclipse lasted only 52 seconds. In any case, ours was a clearly visible event and less fleeting than the literature reports, but limited to just the first component—in other words, no transformation of the spots into bands was seen. This was most likely due to the near equality of the

| ANNULAR ECLIPSES VISIBLE NEAR ITALY THROUGH 2020 | | | | | | |
|---|---|---|---|---|---|---|
| Date | Dura tion | Width | Magni tude | Alti tude | Local Time | Nearby Countries |
| Oct. 3, 2005 | 4:13 | 185 km | 0.951 | 30° | 10:00 | Portugal, **Spain**, Algeria, Tunisia, Libya |
| Jan. 15, 2010 | 8:32 | 351 km | 0.912 | 25° | 08:30 | Uganda, **Kenya**, Somalia |
| Sept. 1, 2016 | 3:05 | 100 km | 0.974 | 70° | 12:00 | Gabon, Congo, Zaire, **Tanzania**, Mozambique |
| Feb. 26, 2017 | 1:08 | 75 km | 0.982 | 14° | 17:30 | Angola |
| Dec. 26, 2019 | 3:02 | 151 km | 0.958 | 11° | 07:30 | Qatar, United Arab Emirates, Saudi Arabia, **Oman** |
| June 21, 2020 | 1:09 | 58 km | 0.984 | 26° | 08:00 | Sudan, **Ethiopia** |

solar and lunar diameters during that eclipse, which produced a more curved-looking solar crescent in the last instants before second contact and the first instants after third contact.

## The Brightness of the Sky

People who have never witnessed a total eclipse often think that total darkness falls during the event. This, in fact, is very rare, at least in our experience. In Mexico in 1991, a rather dark eclipse was expected because the Moon's diameter was much larger than the Sun's, and

therefore, the brighter inner regions of the solar corona would be hidden. It did not happen this way, however. Measurements taken with a camera's light meter showed that the light level dropped by a factor of about 500 during totality as compared with the uneclipsed Sun. In effect, this is the same quantity of sunlight as normally hits the distant planet Uranus. We should remember, nonetheless, that a sky in this state is still at least 1,000 times brighter than the night sky under a full Moon; it is more or less like the sky about a half-hour after the Sun sets.

At the time, we could easily read the shutter speed on our cameras.

The Peruvian eclipse of 1994 was a little darker, more like 45 minutes after sundown, while the Indian one of 1995 was the brightest of all, like the sky after the Sun has just set; indeed, it was not even possible in the short minute of totality to discern the planet Venus. A deeper darkness has surely been recorded during other eclipses. For the most ancient ones, we must turn to the accounts of historical chroniclers and analysts.

*Majestic corona exhibited by the Sun in 1991 Mexican eclipse, photographed in Juchitán on the Isthmus of Tehuantepec by Roberto Crippa and Cesare Guaita.*

*Facing page, bottom, rays of the partially eclipsed Sun penetrate interstices of a shed's roof at San Blas, Mexico, creating images of incredible beauty on the sandy floor. Photograph taken during July 11, 1991, eclipse by Francesco Azzarita of the Barese Astrophiles' Association.*

*Facing page, top, prominences and inner corona during 1994 eclipse, photographed by the author at Gran Chaparral, near Arequipa, Peru.*

We have already referred to the eclipses of 310 and 136 B.C. and to the one in 840 A.D. (see Chapter 7). The *Annales Siculi* reported that during the September 13, 1178, eclipse, "the stars appeared in the sky." Several references to the stars appearing are found in the accounts of another eclipse visible in Italy on June 3, 1239. According to Ristoro D'Arezzo, "night fell" and "all the stars were visible." Giovanni Villani said the same thing: "Night was made of day, and the stars came out." The *Annales Caesenates* is even more explicit: "Almost all the stars were made visible in the sky"; and the Archive of the Cathedral in Siena says, "The stars were as visible as on a clear night."

References to the stars coming out are also found in two Islamic chronicles regarding the April 11, 1176, and May 25, 1267, eclipses that were observed in Turkey; in an account of the May 23, 1221, event observed in Mongolia at the time of Genghis Khan; and in a Chinese journal referring to the eclipse of June 25, 1275.

It is hard to say, however, how accurate these reports may be or how much they may have been influenced by awe and even fear as the events transpired. In other words, were there "many stars" in the absolute sense or only relative to a normal daytime sky in which there are none at all?

A fairly accurate analysis of eclipses over the last two centuries seems to support the second hypothesis. Such an analysis was done in 1972 by Sam Silverman and Gary Mullen. It confirmed that stars of third magnitude were seen only during four eclipses—in 1842, 1860, 1937 and 1940. Among recent eclipses, the only truly dark one, as reported by Bianucci, was the Siberian one of 1981, when stars of second and third magnitude were seen. Since stars brighter than third magnitude can be seen on a night with a full Moon, the popular conception has spread that the darkest totalities are comparable to the light of a full Moon. This statement seems rather overblown considering that on a

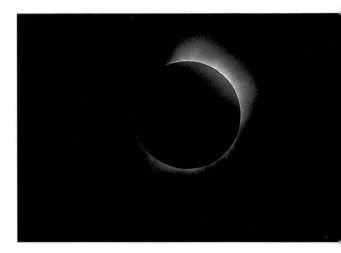

clear night with a full Moon, you can actually pick out stars to fourth and fifth magnitude if the sky is exceptionally transparent (during winter in the mountains, for example). We would maintain that the level of illumination even during a very dark eclipse is at least a few hundred times that of the full Moon.

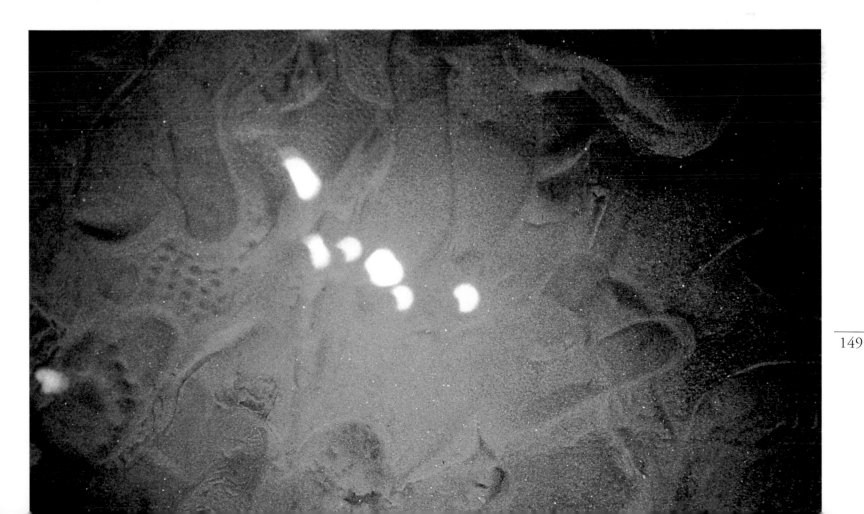

Moon's shadow doesn't make things as dark as we had imagined. No stars other than Sirius are seen, but it is too beautiful all the same, too grand, too unique to find words adequate to describe it . . . and this is all the more true because we had stopped hoping. Someone took pity on us and showed us at least a part of the glory of Nature. Now I know.

The applause that breaks out at the moment of second contact is not a crass display but is, instead, one of the few ways the human soul has to express the fierce emotions produced by this succession of events. The corona can't be seen very well, the outside edge is lost and confused with the clouds, but the inner region is very visible, and its asymmetrical aspect is easy to see. The prominences are spectacular in binoculars; one reaches 130,000 kilometers high. The colors are unreal, the light is something absolutely never seen before; the light level is like an advanced sunset, and the Sun looks like a black hole at the zenith.

Around 360 degrees of horizon, the pink-red glimmer of an impossible sunset hovers; it's the light coming from the areas at the margins of the lunar shadow outside the path of totality. The air temperature has fallen from 40 to 29 degrees C. The birds had begun flying low during the partial phases, but they started behaving normally again when totality began. As second contact approached, a few chickens in a nearby courtyard headed for the chicken coop and have now ceased every type of movement.

*El Gran Chaparral, Pampa de La Joya, 16°42'S, 71°54'W, Arequipa Province, Peru, November 3, 1994.*

Alarm rings at 2:30 a.m. Departure from Arequipa in the most dismal darkness, loaded to the hilt with telescopes, binoculars, cameras and video cameras. During the trip, the coming

## A Travel Journal

Technical terms are inadequate to convey what a total eclipse of the Sun looks like to someone who has never observed one. So it may be preferable to let the emotions speak by presenting three short eyewitness accounts written by the author.

*San Blas, 21°31'N, 105°16'W, Nayarit State, Mexico, July 11, 1991.*

A sudden hope. The sea breeze arrests a cloud and then takes less than five minutes to rip it right in half and part it like a comb through hair, revealing the Sun in the middle. It doesn't last more than a split second, though. Two minutes later, with the sudden cooling of the atmosphere (the Sun is about to disappear), the wind drops and the cloud takes the

upper hand. But the Sun doesn't disappear, and the cloud isn't so black. Someone removes his telescope's solar filter a little early and risks momentary blindness, which means that there is still a lot of light, the cloud cover is thin enough and all is not lost.

When no one dares hope any longer, the scene changes, a magical halo around the Sun appears, well-defined fuchsia-colored prominences are clearly silhouetted in binoculars— one can even be seen with the naked eye. Someone took the Sun away, and there's a black spot in its place, right in the middle of the sky, in the middle of the world. It's not what we had imagined; it's different from everything we had envisioned, but no less beautiful. No one sees any shadow bands, and the advance of the

*Solar corona photographed during 1994 eclipse by Roberto Crippa and Cesare Guaita in Sevaruyo, Bolivia, at 3,800 meters.*

and going of clouds and stars; through the car windows, some pick out the Southern Cross and Alpha and Beta Centauri. Finally, at 4:15, El Gran Chaparral, a real desert with a kind of green oasis nearby and the Arequipa mountains very far away.

At 5:30, the sunrise surprises us, but our star soon disappears behind a bank of clouds. The first of the partial phases, which begin at 6:06, are invisible; the cloud bank is huge, the cloud cover thick. Misery begins to sink into our hearts. If this continues, we'll see the dark fall, but that will be the end of it! Around 6:40, a brief clearing doesn't nourish much hope, because the clouds immediately devour the Sun again.

Our morale has fallen beneath our shoes when, miraculously, at 7:05, 10 minutes before totality, the clouds clear away decisively, and the solar photosphere reappears. I get ready to shoot totality with high expectations. Everyone around me is frenetic. Everyone had given up hope of seeing anything, so this is a gift from the sky. The drop in light level in the last five minutes is impressive. The clouds take on surreal reds and oranges. At 7:17, the Moon's black shutter switches off the Sun's last spark. The corona appears, and I begin to tremble like a leaf, maybe because it's cold, maybe with emotion, I don't know. I take a few shots at the beginning, and then for a long minute, I just relish the eclipse naked-eye and through binoculars.

I look around me. It isn't pitch-dark, but it is a little darker than it was in Mexico—like three-quarters of an hour after sunset—and the clouds are improbably colored. In this light, the landscape seems flattened. As Franca Pontecorvo tells me later, the Arequipa volcanoes seemed to move immeasurably closer and were standing there looking on like the rest of us.

I watch the sky where the Sun reposes; the

white corona is fixed in a sky that seems clear, because it is cobalt blue. Surely the clouds can't be very dense; the sky is arrayed in layers of cloud alternating with blue spaces. Above and to the right, the splendid glow of Venus adorns the scene with enchantment (Marino Vago even sees Mercury, but Jupiter stays behind a thick cloud for the duration). I look at the corona through binoculars and see some distinct lateral protrusions, but no prominences.

When it seems the eclipse is about to end, I return to my camera. It ends in glory. I have just enough time for two shots, when I see the chromosphere through the camera. I take the shots while instinctively glancing at the Sun in the same moment. A fuchsia, then blood-red line catches fire on the upper left rim and drags a shout from the bottom of my being: "The chromosphere!" The line explodes and becomes a bubble, first red, then white. A gigantic, brilliant diamond ring flares forth to signal the end of the event. It leaves everyone gaping in wonder. *Sariska Wild Life Sanctuary, 27°27'N, 76°33'E, State of Rajasthan, India, October 24, 1995.*

We have stationed ourselves close to the road, and some curious people begin to stop

*Above, exceptional diamond ring seen at the end of totality during the 1994 eclipse was "contaminated" by red light from the chromosphere and by passing clouds. Photograph taken at Gran Chaparral by Gianvittore Delaito.*

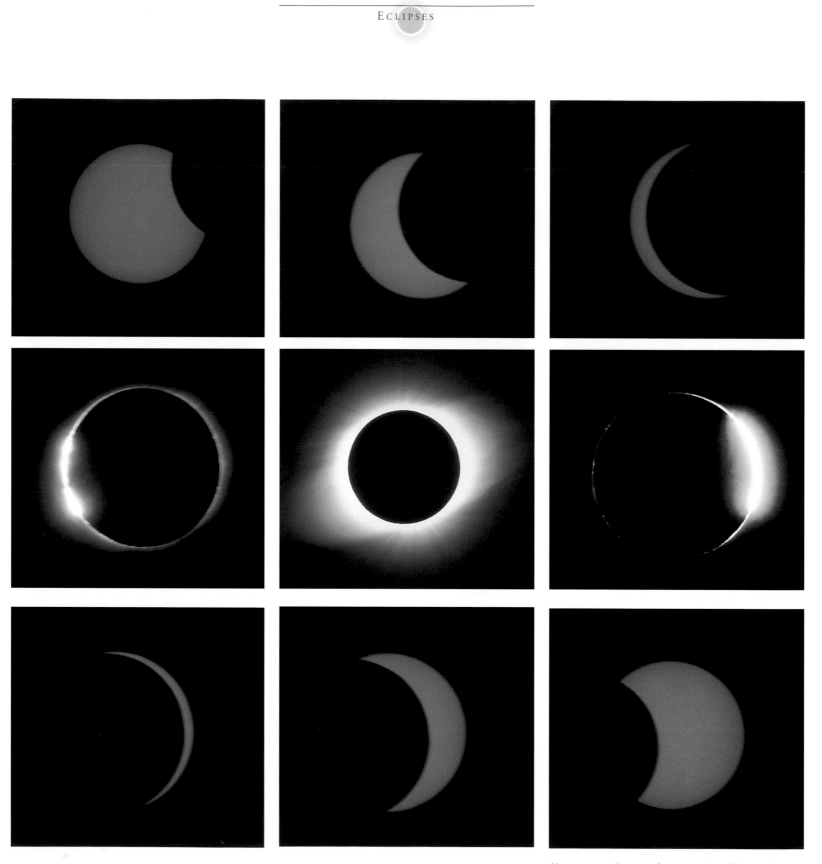

*Above, sequential images of October 24, 1995, eclipse taken by the author in India. From left to right and top to bottom: (a) magnitude 0.23; (b) 0.66; (c) 0.90; (d) double diamond ring at second contact; (e) corona at maximum extension; (f) diamond ring at third contact; (g) magnitude 0.94; (h) 0.77; (i) 0.42.*

by. We give them a couple of peeks through our binoculars and camera lenses. We exhort our driver, Daulad, to keep people far away from us at the moment of totality, from 8:33 to 8:34 a.m. I explain that this is the most important moment for us and that it would be foolish to have it ruined by a banality after such a long journey. I also explain briefly what is about to happen; he seems very interested, maybe even a little concerned. The moment approaches. The sky stays completely clear. The temperature starts to drop. I put on my sweater. About 8:15, the drop in illumination is now remarkable. Once he has learned how they work, Daulad takes pleasure in watching the eclipse through my binoculars and roams about happily with them hanging over his shoulder.

Five minutes before the witching hour, I have to remind him to give them back to me! And here it is. It's time to get very serious. It's time to really buckle down. I'm agitated, as usual, and I try to concentrate to overcome the tremor that rocks me. I can't do anything dumb. Unlike other times, the Moon's progress seems very slow to me; this is because the Moon's apparent diameter is slightly bigger than the Sun's. Through my camera's viewfinder, I see the chromosphere appear up high, while the photosphere down low has not yet disappeared. I start the timer, probably a little ahead of time. I go with short exposures for the chromosphere and prominences. Everything is so brief, so feverishly frantic. It seems I don't have enough time to get anything done. I switch the camera to the "B" setting for long exposures of the corona—a half-second, a second, two seconds. I adjust the shutter speed back to 1/125, ready to catch another diamond ring in the final moments. Finally, I watch. The corona presents two unusually long equatorial plumes that pro-

ject almost vertically, due to the Sun's orientation on the horizon. Each one must be at least three solar diameters long, maybe more. I raise the binoculars to attempt a better view; I see the chromosphere, catch a glimpse of a few prominences, but there is no time! My God, there is no time. The timer goes off, but it isn't over yet. Another few seconds, and a light kindles at the top edge of the Sun. It grows, becomes gigantic, explodes and just won't stop. It is a colossal diamond ring that lasts at least four or five seconds. I try to capture it on film, all the while awestruck by the apparition.

The reappearance of the photosphere ends my perilous apnea. We have definitely just experienced the most hyper moment of our lives. I sense that we were given a few extra seconds of totality, but I still don't feel that I did anything right, whether watching or taking pictures. It was truly touch and go. I look at Gianvittore. He has his mouth open too, overcome by emotion and the brevity of it all. We only now realize there is no one around us. Atavistic terror? As soon as the photosphere reappears, however, we see our Daulad leap out of the car, together with a passerby who had stopped, and ask, "Is it over, really, is it over?" Evidently, the phenomenon did not leave him indifferent!

153

*This image shows a different kind of total eclipse of the Sun, with the Earth hiding our star, taken by Apollo astronauts on November 12, 1969. (NASA)*

the path of totality for the August week centered on the eclipse. He found that only 1 percent of the days were without sunshine. In general, over two-thirds of the data recorded more than 10 hours of sunshine every day. Among all Hungarian sites, Lake Balaton seems to be the most favorable. It is yet another tourist destination of great charm in summer, characterized by warm waters and sandy beaches. The lake also carries a considerable historic interest, inasmuch as it marked the border between the Ottoman and Hapsburg empires in the 16th and 17th centuries.

All of eastern Europe is protected by the Carpathian and Balkan mountain chains and, during the summer, is under the influence of a Mediterranean climate, flowing from the Adriatic Sea. Farther west, Atlantic currents prevail, and the situation is less rosy. The probability drops to 53 percent south of Vienna and 51 per-

cent in Monaco. Still farther west, the situation becomes even more negative, and we would exclude western Germany and France as travel destinations for the 1999 eclipse. According to Espenak and Anderson, the best conditions in Germany will be found between Ulm and Munich and in Austria, southeast of Vienna along the Hungarian border. Anyone thinking about going to Austria or Germany should obviously keep an eye on the weather in the preceding days. For example, the development of the Azores anticyclone (a weather formation whose evolution has really been rather irregular in the past few years) during the time of the eclipse would have a stabilizing effect on the weather throughout central Europe. This, together with the fact that the eclipse will clearly happen before the daily peak of cloud formation, may render a view of the eclipse a near certainty. Because their required travel and

stopover time will be so brief, even people who leave from Italy can check weather reports relevant to cloud cover up to the last minute (via the Internet or from satellite images transmitted by various media) and then set out for the most promising destination.

The best thing would be to arrive at a favorable area ahead of time, perhaps the evening before, and then, with the help of a portable computer linked to the Internet, to refine your location before the first light of day. Stay informed about the weather for the various places involved in the event as well as German and Austrian border crossings. The importance of the event will not be lost on Europe's mass media. Thinking back to Comet Hale-Bopp will give you a good idea of what can happen. Millions of people mobilized themselves to see that comet, and millions will probably want to see the eclipse. Hale-Bopp was visible for a few

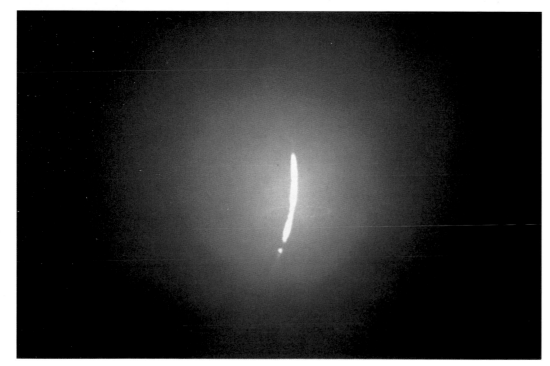

weeks, however, and there is only one eclipse day. It might not be very bold to predict a real exodus from the south and north toward the center of Europe. Transportation workers in these countries will more than likely be overwhelmed, so don't take chances yourself—choose alternative routes wherever possible instead of the usual mountain passes.

An important consideration when choosing

*Left, Baily's beads photographed by the author just before second contact during 1991 eclipse using a 1,000mm f/10 lens.*

*Above, images of solar corona obtained with lenses of varying focal lengths.*

an observing location involves your distance from the center line. Contrary to what you might think, the duration of the eclipse does not diminish all that much until you get a long way from the line. In this eclipse, for example, at 10 kilometers from the center line, you lose only 2 seconds; at 15 kilometers, 5 seconds; and at 20 kilometers, 9 seconds. Even at 30 kilometers from the center line, there are still 2 minutes of totality. This has to be kept in mind in case, for various reasons (better terrain, more accessible services, special compositions for your photographs, etc.), you consider staying at some distance from the center line.

The table on page 157 will help you quickly determine the beginning and end times of totality at the place chosen for observing. You should note that first and fourth contacts are separated on the center line by a minimum of 2 hours 45 minutes in Germany and a maximum of 2 hours 47 minutes in Romania and that, as this is being written, the same local time—two hours ahead of Greenwich time—is in effect in August in every country under discussion.

## Observing the Eclipse

We could be witty and say that the only truly indispensable tool among all the instruments you might carry to see an eclipse is your eye. Once again, its great capacity as an exceptionally sensitive and versatile instrument is confirmed. Only the naked eye allows us to observe so much: our surroundings, the colors of the ground and sky, the true splendor of the corona and the stars that may ultimately be visible.

During the partial phases, it is imperative that the eye be protected by a rather strong, dark filter. One safe and affordable filter is the common

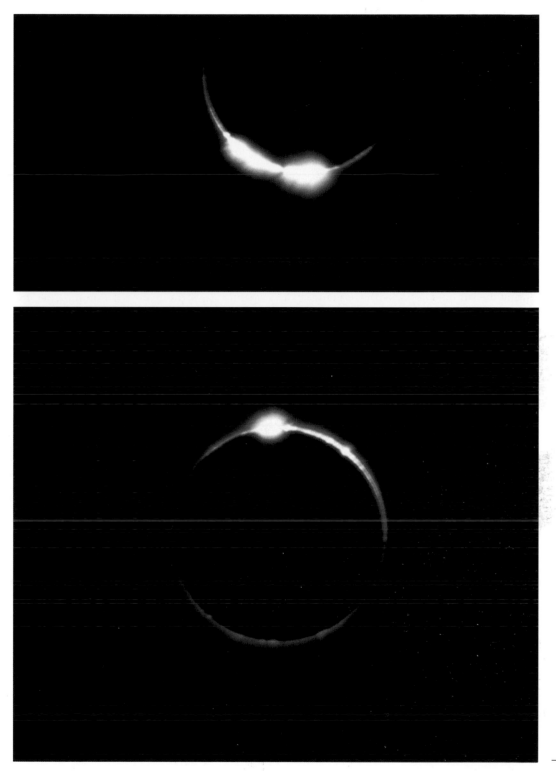

*Diamond rings at beginning (top) and end (bottom) of totality during 1995 eclipse in India photographed by Gianvittore Delaito with a 500mm f/8 lens.*

159

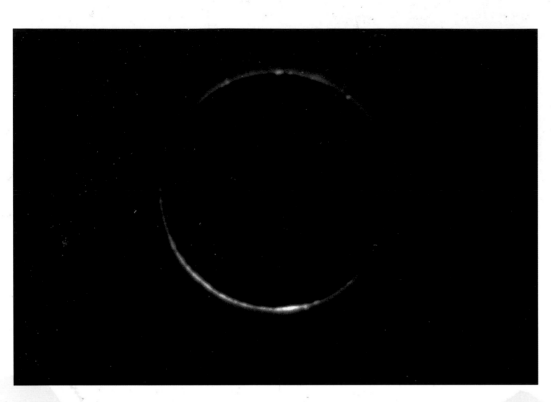

furnish. Binoculars are unbeatable for observing the outer corona and are certainly adequate for observing Baily's beads, the chromosphere, prominences, sunspots and coronal detail. As with comets, 20x80 binoculars are probably the maximum a neophyte can use for visual observation of a total eclipse. Again, the filters must be removed during totality, but take care not to remove them before second contact. (The same is true if using a telescope.) Beyond potentially damaging the retina, the blaze transmitted from even a sliver of the still visible photosphere would certainly impede your view of the fainter corona.

## A Lifelong Memory

Viewing an eclipse firsthand is obviously the main thing. But people want a lasting souvenir of an event they are likely to see only once. Many years afterward, it is hard to recall, but a good photograph revives everyone's memory.

As a starting point, you can even shoot the eclipse with a simple little automatic camera. Naturally, the low light level during totality will make the flash go off automatically (how many thousands of flashes went off during the 1991 eclipse!), so the camera should permit you to disconnect the flash and select the shutter speed manually. With a film speed of 100 ISO and the lens at f/3.5, this can be 1/250 to 1/30 second during totality, because the corona, even if it is very faint, will always equal in luminosity the full Moon or a landscape illuminated by the Sun. At the slower shutter speed, the corona will appear at its maximum extension.

To take more meaningful photos, of course, you need a reflex camera. A 200mm lens seems to be the minimum to ensure enough image area for the solar corona. At its average exten-

#14 welder's glass, which gives the Sun a green cast. Alternatively, you can use a Mylar filter (often mounted in special eclipse-viewing glasses made expressly for this purpose), sold by many telescope manufacturers and dealers as well as science stores; it gives the Sun a bluish hue. Sunglasses, floppy disks or CDs, smoked glass, photographic film (exposed or not), polarizing filters, neutral photographic filters and x-ray film are absolutely *not* safe. All of them, to some degree—especially in the case of prolonged use—can cause temporary or permanent damage to the retina. A proper filter must be used continuously, without fail, right through to the moment of second contact and immediately after third contact (even 1.1 percent of the photosphere can be fatal to the exposed eye). The period of totality, on the other hand, not only *can* but *must* be observed without filters;

otherwise, you risk missing the whole show.

Many people own binoculars, another precious instrument (if properly filtered until totality). Any binoculars at all, even opera glasses, are good for observing the phenomenon. But here again, you *must* use filters during all the partial phases. To dim the incoming sunlight adequately, the filters must be mounted on the lenses (not on the eyepieces, where they might shatter). Any optical-supply store or photographic equipment dealer should be able to procure proper filters. A pair of 7x50, 10x50 or 12x50 binoculars is good for observing an eclipse; in some ways, binoculars are even superior to a telescope. Of course, a telescope offers a better view of the prominences, the chromosphere and the fine details of the corona, but binoculars encompass an image of the entirety of the spectacle that no telescope can

*Chromosphere photographed in India by Gianvittore Delaito with a 500mm lens.*

*Facing page, diamond ring photographed by Roberto Crippa and Cesare Guaita in Bolivia on November 3, 1994, with an 800mm f/8 refractor.*

sure times between 1/60 and 1/8 should be avoided, because they will yield blurry images due to the vibration produced when the camera's mirror lifts (unless the camera has a manual lifting device). Times longer than 1/8 can be held manually using the method discussed in Chapter 9—setting the shutter on "B," using a cable release and keeping the lens covered with cardboard or a lens cap. After three or four seconds, when the vibrations subside, remove the cardboard, then replace it after 1/4, 1/2 or 1 second, and so on. Even though the exposure times are approximate with this system, the results are excellent.

Obviously, these remarks are valid when the sky is clear. If the sky is hazy or cloudy, increase the aperture by at least two or three f-stops. Since clouds cause a uniform diffusion of light, you can rely on the camera's light meter.

You can also try shooting the entire sequence of the eclipse on a single frame of film using the same method laid out in Chapter 9 for a lunar eclipse (following the instructions given about f-stops, shutter speeds and film). You will have to time the exposures manually at 1/4 second for the entire sequence (including totality). Just at totality, open the shutter completely and remove the filter without moving the camera.

Another exciting shot to attempt is a wide-field image of the eclipsed Sun and any bright planets and stars nearby. Only two planets will be above the horizon during the eclipse of August 11, 1999. Venus, at magnitude –3.5, and Mercury, at magnitude 0.7, will be found 15 degrees east and 18 degrees west of the Sun, respectively. Among the brightest stars near the Sun that can be photographed will be Regulus (magnitude 1.3), at 10 degrees east; Castor and

*The planets Venus (above the Sun) and Jupiter (below) appear in this wide-field image of the November 3, 1994, eclipse taken by Serge Brunier In the Chilean Andes. The Pomerape volcano is visible at bottom.*

Pollux (1.9 and 1.1), at 31 degrees and 28 degrees northwest; and Procyon (0.4), at 30 degrees southwest. With a 35mm lens, 100 ISO film and a wide aperture, you can attempt exposures of 2, 4, 8 and even 16 seconds to capture the stars together with the eclipsed Sun. The longer exposures will usually reveal planets and stars that are not visible to the naked eye during the event.

It is also worth paying attention to the chance appearance of shadow bands anytime from three minutes after third contact. Good photographs of this phenomenon are very rare. Since the shadows move rapidly and are of extremely low contrast, the experts advise using a normal 50mm lens, a sensitive film around 400 ISO and a fast shutter speed—say 1/250 second. Considering the rarity of this phenomenon, however, we recommend trying several differ-

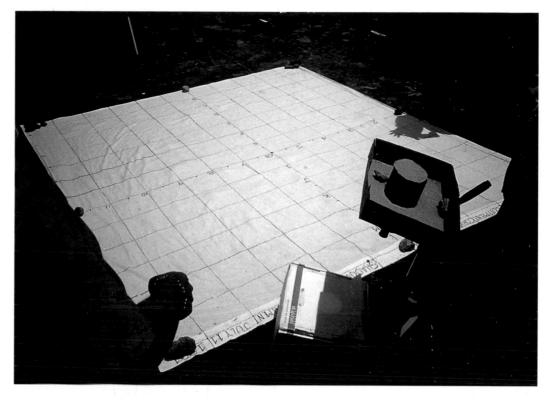

ent apertures and shutter speeds and taking several shots. To improve visibility and contrast, it is helpful to spread a white bedsheet on the ground.

The eclipse can also be shot with a video camera since, among all the celestial phenomena we've talked about, it is the only one bright enough to give off adequate light. Once again, it is absolutely essential to use a filter for the partial phases.

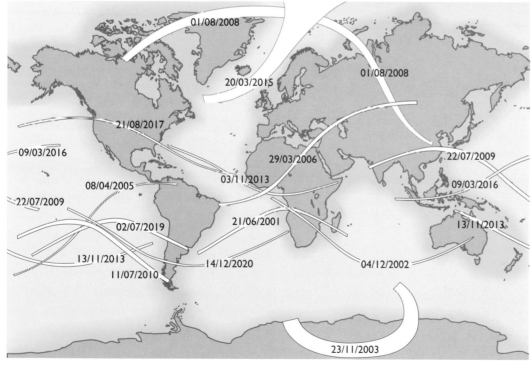

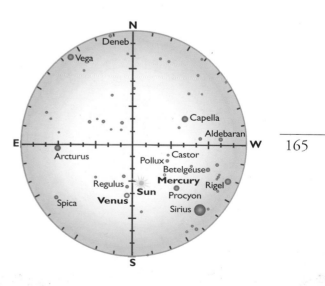

*Top, a white bedsheet spread out on the ground by American amateur astronomers at San Blas, Mexico, to observe and photograph shadow bands during the 1991 eclipse. Sheet is subdivided into equal units to measure the speed of the shadows. Photograph by Francesco Azzarita.*

*Above, paths of totality for eclipses listed in lower table on page 157.*

*Right, map of the sky for August 11, 1999, eclipse, showing positions of the planets and brightest stars.*

# Bibliography

**The Great Comets**
Apianus, Petrus. *Astronomicum Caesareum*, Ingolstadt, 1568.
Böhm, Conrad; Fulle, Marco. "Una cometa per Napoleone," *L'Astr.*, Jul. 1996.
Bortle, John E. "Comets and How to Hunt Them," *S&T*, Feb. 1981.
Bortle, John E. "Great Comets in History," *S&T*, Jan. 1997.
Celoria, Giovanni. *Le Comete*, Treves, Milan, 1873.
Di Cicco, Dennis; Robinson, Leif J. "Comet Photography for Everyone," *S&T*, May 1996.
Fulle, Marco. "Dalla coda all visibilità di una cometa," *Astronomia, UAI*, Jan./Mar. 1981.
Hasegawa, Ichiro. "Catalogue of Ancient and Naked-Eye Comets," *VIA*, Vol. 24, 1980.
Hevelius, Johannes. *Cometographia*, Danzig, 1668.
Hughes, David H. "The Criteria for Cometary Remarkability," *VIA*, Vol. 30, 1980.
Kronk, Gary W. *Comets: A Descriptive Catalog*, Hillside, New Jersey, Enslow, 1984.
Kronk, Gary W. "The Great Comet of 1811," *ICQ*, Jan. 1996.
Lubienietz, Stanislaus. *Theatrum cometicum*, Amsterdam, 1681.
Luu, Jane X.; Hewitt, David C. "La fascia di Kuiper," *Le Scienze*, Sept. 1996.
Maffei, Paolo. *La cometa di Halley*, Mondadori, Milan, 1984.
Martin, Franco Foresta. *Le Comete*, Sansoni, Florence, 1982.
Needham, Joseph; Beer, Arthur; Yoke, Ho Peng. "Spiked Comets in Ancient China," *Observatory*, Vol. 77, 1957.
Newburn, R.L.; Neugebauer, M.; Rahe, J. (eds.). *Comets in the Post-Halley Era* (2 volumes), Kluwer Acad. Publ., Dordrecht, 1991.
Pingré, Guy Alexander. *Cometographie*, Vol. 1, Paris, 1783.
Rodgers, R.F. "Newly Discovered Byzantine Records of Comets," *The Journal*, The Royal Astronomical Society of Canada, Vol. 46, No. 5, 1952.
Romano, Giuliano. "Stelle e comete nell'astronomia precolombiana," *Coelum*, May/June 1986.
Schiaparelli, Giovanni Virginio. "La cometa" and "La cometa del 1882," in *Le più belle pagine di astronomia popolare*, Hoepli, Milan, 1925.
Stephenson, Richard F. "Osservazioni celesti nelle cronache islamiche," *L'Astr.*, Apr. 1993.
Stern, Alan. "Where Has Pluto's Family Gone?" *Astronomy*, Sept. 1992.
Stern, Alan; Campins, Umberto. "Chiron and the Centaurs: Escapes From the Kuiper Belt," *Nature*, Aug. 8, 1996.
Tempesti, Piero. *I segreti delle comete*, Curcio, Rome, 1984.
Vanin, Gabriele. "Le grandi comete," *L'Astr.*, Apr. 1994.
Vanin, Gabriele. *La cometa Hale-Bopp: come, dove, quando osservarla*, Pilotto, Feltre, 1996.
Vanin, Gabriele. "Comete, angoli e strumenti," *L'Astr.*, Feb. 1997.
Various Authors. *Comete, asteroide, meteoriti*, Le Scienze notebooks, No. 26, 1985.
Various Authors. "Tabulation of Comet Observations," *ICQ*, Vols. 15–18 (1993–96).
Vsekhsvyatskij, Sergey Konstantinovich. *Physical Characteristics of Comets*, Israel Program for Scientific Translation, Jerusalem, 1964.
Weidenshilling, S.J. "Origin of Cometary Nuclei as 'Rubble Piles,'" *Nature*, Apr. 21, 1994.

Weissman, Paul. "Bodies at the Brink," *The Planetary Report*, Jan./Feb. 1994.
Whipple, Fred. "La natura delle comete," *Le Scienze*, May 1974.
Wilkening, Laurel L. (ed.). *Comets*, University of Arizona Press, Tucson, 1982.
Yeomans, Donald K. *Comets: A Chronological History of Observation, Science, Myth, and Folklore*, Wiley, New York, 1991.
Yeomans, Donald K. "Comets: Historical Apparitions," in *The Encyclopedia of Astronomy & Astrophysics*, Van Nostrand Reinhold and Cambridge University Press, New York and Cambridge, 1992.
Yoke, Ho Peng. "Ancient and Mediaeval Observations of Comets and Novae in Chinese Sources," *VIA*, Vol. 5, 1962.

**Meteor Showers**
Bone, Neil. *Meteors*, Sky Publishing Corp., Cambridge, 1993.
Cecchini, Gino. *Il cielo*, UTET, Turin, 1969.
Coco, Mark J. "Catch the Geminid Meteor Shower," *Astronomy*, Dec. 1993.
Croswell, Ken. "Will the Lion Roar Again?" *Astronomy*, Nov. 1991.
Di Cicco, Dennis. "Photographing the Perseids," *S&T*, Aug. 1993.
Hughes, David W. "The History of Meteors and Meteor Showers," *VIA*, Vol. 26, 1982.
Hughes, David W. "A Mysterious Woodcut," *S&T*, Sept. 1987
Hughes, David W. "The Life and Death of a Meteoroid Stream," *S&T*, Aug. 1993.
Jacchia, Luigi G. "Cade una stella," *L'Astr.*, Jul./Aug. 1981.
Littmann, Mark. "The Discovery of the Perseid Meteors," *S&T*, Aug. 1996.
MacRobert, Alan M. "Meteor Observing I" and "Meteor Observing II," *S&T*, Aug. and Sept., 1988.
Olson, Donald W.; Olson, Marilynn S. "William Blake and August's Fiery Meteors," *S&T*, Aug. 1989.
Olson, Donald W.; Doescher, Russell L. "August Meteors in the 1860s," *S&T*, Aug. 1993.
Rao, Joe. "Storm Watch for the Perseids," *S&T*, Aug. 1993.
Rao, Joe. "The Leonids: King of the Meteor Showers," *S&T*, Nov. 1995.
Schiaparelli, Giovanni Virginio. "La pioggia delle stelle," in *Le più belle pagine di astronomia popolare*, Hoepli, Milan, 1925.
Sinnott, Roger W. "The Optimum Camera for Meteors," *S&T*, Feb. 1994.
Vanin, Gabriele. *Stelle cadenti*, Galliera V., Biroma, 1994.

**Eclipses**
Abetti, Giorgio. *Esplorazione dell'universo*, Laterza, Bari, 1965.
Akasofu, Syun-Ichi. "The Shape of the Solar Corona," *S&T*, Nov. 1994.
Anderson, Jay; Willcox, Ken. "Return to Darkness," *Astronomy*, Nov. 1993.
Aveni, Anthony. *Gli imperi del tempo*, Dedalo, Bari, 1993.
Baroni, Sandro. "Tre eclissi totali di Sole per l'Italia settentrionale," *Coelum*, Nov./Dec. 1986.

Bianucci, Piero. "In Siberia per l'eclisse di Sole," *L'Astr.*, Nov./Dec. 1981.
Bianucci, Piero. *Rapporto sul Sole*, Rusconi, Milan, 1982.
Codona, Johanan. "The Enigma of the Shadow Bands," *S&T*, May 1991.
Dickinson, Terence. *NightWatch*, Firefly, Willowdale, Ontario, and Buffalo, New York, 1998.
Espenak, Fred. *Fifty Year Canon of Solar Eclipses: 1986–2035*, Sky Publishing Corp., Cambridge, 1987.
Espenak, Fred; Anderson, Jay. *Total Solar Eclipse of August 11, 1999*, NASA Ref. Publ. 1398, Greenbelt, 1997.
Godoli, Giovanni. *Il Sole*, Einaudi, Turin, 1982.
Goldreich, Peter. "Le maree e il sistema Terra-Luna," *Le Scienze*, Jul. 1972.
Herodotus. *Le storie* (*Histories*), Mondadori, Milan, 1988.
Littmann, Mark; Luu, Jane; Marsden, Brian; Jewitt, David; Trujillo, Chadwick; Hergenrother, Carl; Chen, Jun; Offutt, Warren. "A New Dynamical Class of Object in the Outer Solar System," *Nature*, June 5, 1997.
Neugebauer, Otto. *Le scienze esatte nell'antichità*, Feltrinelli, Milan, 1974.
O'Meara, Stephen James. "Making Sense of November's Perplexing Lunar Eclipse," *S&T*, June 1994.
Pasachoff, Jay M. "La Corona Solare," *Le Scienze*, Jan. 1974.
Proverbio, Edoardo. *Archeoastronomia*, Teti, Milan, 1989.
Rao, Joe. *Your Guide to the Great Solar Eclipse of 1991*, Sky Publishing Corp., Cambridge, 1989.
Reynolds, Michael D.; Sweetsir, Richard A. *Observe Eclipses*, Washington, The Astronomical League, 1995.
Secchi, Angelo. "Lo spettacolo di un'eclisse," in *Astronomia alla scoperta del cielo*, Curcio, Rome, 1983.
Software Bisque. *The Sky*, Golden, 1992.
Stephenson, F. Richard. "Eclissi storiche," *Le Scienze*, Dec. 1982.
Stephenson, F. Richard. "Antiche eclissi e i ritardi dell'orologio-Terra," *L'Astr.*, May 1983.
Stephenson, F. Richard. "Antiche eclissi italiane," *L'Astr.*, Feb. 1987.
Stephenson, F. Richard. "Gli astronomi arabi e le eclissi," *L'Astr.*, Apr. 1992.
Vanin, Gabriele. "9 febbraio: la Luna in passerella," *L'Astr.*, June 1990.
Vanin, Gabriele. "UAI Maya Tour," *Astronomia, UAI*, Sept./Oct. 1991.
Vanin, Gabriele. "L'eclisse del 24 ottobre 1995 e la forma della corona solare," *Astronomia, UAI*, Nov./Dec. 1995.
Vanin, Gabriele. "Breve ma intensa l'eclisse indiana," *L'Astr.*, Feb. 1996.
Vanin, Gabriele. "Quando si nasconde la Luna," *L'Astr.*, Apr. 1996.
Vecellio, Antonio. *Storia di Feltre*, Vol. IV, Castaldi Feltre, 1877.
Willcox, Ken. *Totality: Eclipses of the Sun*, University of Hawaii Press, Honolulu, 1991.

Abbreviations:
*ICQ* = *International Comet Quarterly*, Cambridge, Mass.
*L'Astr.* = *L'Astronomia*, Milan
*S&T* = *Sky & Telescope*, Cambridge, Mass.
*UAI* = *Unione Astrofili Italiani*, Rome
*VIA* = *Vistas in Astronomy*, London

# Index to Illustrations